ASIATIC PRIMULAS

P. bracteosa (Sect. Petiolaris) ̈ ($\times \frac{1}{2}$) (p. 63)

ASIATIC PRIMULAS

A Gardeners' Guide

by

ROY GREEN

THE ALPINE GARDEN SOCIETY

General Editor: R. C. Elliott VMH

For the members of
THE ALPINE GARDEN SOCIETY
May they, and their primulas, flourish

Edited by
Dr. Lionel J. Bacon

© The Alpine Garden Society 1976
First published Autumn 1976 by
The Alpine Garden Society
Lye End Link, St. John's, Woking, Surrey
Reprinted 1978
by L. Baker (Printers) Ltd.
71 Lombard Street, Birmingham B12 0QU

CONTENTS

LIST OF ILLUSTRATIONS

LINE DRAWINGS

PHOTOGRAPHS

FOREWORD

I cannot remember the time when I first became interested in Asiatic primulas. The knowledge that they grow on the great mountains of central and south-central Asia, the lonely places, would certainly induce a sense of nostalgia. Many species are very garden-worthy plants of unsurpassed beauty, graceful in habit, with flowers of delicate form and restrained colour. Many species are ideally woodland plants and, while there is no woodland about my garden, there is certainly a coolness which creates a near-substitute in buoyancy.

When I was asked to produce this monograph I was rather horrified, for I enjoy my primulas, which for one thing means that I have had neither time nor inclination to pry into their intimate botanical minutiae.

It was however felt that for this monograph to serve its purpose adequately, each species described should be identifiable from the notes given. This, of course, required the use of some botanical phraseology and availability of the many dimensions. For this information, reference has been made almost entirely to the standard work of Smith & Fletcher (5), (6) and (7).*

Sir William Wright Smith and Dr. H. R. Fletcher were, each in his turn, Regius Keepers of the Royal Botanic Garden, Edinburgh, and this was at a period when more species of Asiatic primula were in cultivation than either before or since. They also had access to the great herbaria of the world. Smith and Fletcher's work was published over a period of eight years, and based on more than twenty years of professional observation. For the assistance of anybody who wishes to have greater detail regarding any particular species, the references of the various sections in which the standard work was published are given in the Appendix.

I have also to thank Mr. Alfred Evans of the Royal Botanic Garden, Edinburgh and President of the Scottish Rock Garden Club for his time and much kindness, and Dr. John Richards of Hexham for assistance with that tricky group, the Petiolarids; and the Regius Keeper of the Royal Botanic Garden, Edinburgh for making available a number of the photographs.

Several of the photographs are reproduced from the A.G.S. *Bulletin*, and thanks are due to the Editor, Mr. R. C. Elliott, VMH,

* The numbers in brackets relate to the references on p. 148.

for making this possible, as well as for a number of other photographs of his own.

I have to thank Mr. A. Duguid of Coldingham, Berwickshire, for information on the *Primula strumosa* hybrids, and Mr. J. C. Lawson of Aviemore, Inverness-shire for information on the section Minutissimae.

The series of booklets on Asiatic Primulas published many years ago by the firm of W. E. Th. Ingwersen Ltd. proved a most valuable aid in compiling the index, and I am indebted to Messrs. Will and Paul Ingwersen for making them available.

There is also a 'List of Primula Species in Cultivation' by Mr. R. Edwards and Dr. D. Winstanley. This list includes European as well as Asiatic primulas: I am grateful to have had the opportunity to refer to the draft of this paper.

Finally, I have to thank my editor, Dr. Lionel J. Bacon, who is a man of tremendous energy and perception. Dr. Bacon produced the keys to the various sections and relieved me of the tremendous burden of compiling the glossary and the general index. He has also proved to be a most excellent illustrator.

INTRODUCTION

As the work of discovery, identification and description of the various species of *Primula* continued to develop during the ninteenth and earlier part of the twentieth century, it became evident that a very large number of species was involved and that some form of grouping was necessary in order to comprehend the genus as a whole.

The first attempt at a systematic classification of the genus *Primula* is probably that due to Pax (2). This was followed by a paper on the Chinese species of primula given by Balfour at the Third Primula Conference (3) and a major revision by Smith and Forrest in 1928 (4).

Even at the time of the Fourth Primula Conference in 1928 (4) the significance of the morphology and geographical distribution of various affinities was not clearly understood. These difficulties were, however, largely resolved, in part by a consideration of cytological data, subsequent to the publication of a series of monographs by Smith and Fletcher during the years 1942–48, (5), (6) and (7). Although there have been a few minor additions and alterations since 1948, Smith and Fletcher's work remains a very thorough, detailed and reasoned appraisal of the genus, and it is upon their system of classification that the present work is based.

Although Smith and Fletcher discuss the relationship between the various groups of *Primula* in considerable detail, their monographs are distributed in the Proceedings of three separate learned societies and no sequential list of sections was published. We have, therefore, to return to Smith and Forrest's monograph of 1928 (4), upon which the following summary of sections is based.

The relationship between the various sections is much the same in the monograph of 1928 and the later series, although the species attributed to the various sections were varied slightly in the light of additional information and analysis.

The following Table therefore lists the sections under which the Asiatic species of the genus will be considered. The sections Megaseaefolia, Bella, and Verticillata of the 1928 monograph are now included in the sections Vernales, Minutissimae, and Floribundae respectively. Species of the section Obtusifolia have been distributed between the sections Farinosae and Nivales, and the new section Parryi has been created to accommodate the North American Nivalid relatives.

[1]

1.	CORTUSOIDES Balf.f.	17.	MINUTISSIMAE Pax
2.	REINII Balf.f.	18.	SOULIEI Balf.f.
3.	PINNATAE R.Knuth	19.	FARINOSAE Pax
4.	PYCNOLOBA Balf.f.	20.	DENTICULATA Balf.f.
5.	OBCONICA Balf.f.	21.	CAPITATAE Pax
6.	SINENSES Pax	22.	MUSCARIOIDES Balf.f.
7.	MALVACEA Balf.f.	23.	SOLDANELLOIDEAE Pax
8.	CAROLINELLA Pax	24.	ROTUNDIFOLIA Balf.f.
9.	VERNALES Pax (European)	25.	NIVALES Pax
10.	AMETHYSTINA Balf.f.	26.	PARRYI W.W.Sm. (North
11.	CUNEIFOLIA Balf.f.		American)
12.	PETIOLARES Pax	27.	SIKKIMENSIS Balf.f.
13.	MALACOIDES Balf.f.	28.	CANDELABRA Balf.f.
14.	GRANDIS Balf.f.	29.	AURICULA Pax (European)
15.	BULLATAE Pax	30.	FLORIBUNDAE Pax
16.	DRYADIFOLIA Balf.f.		

In the description of species which follows, species have been listed section by section, and sections treated in the sequence of the preceding table.

The section lists are, as far as is known, complete with respect to the Asiatic species, with the exception of a very few species which have perhaps only been found on a single occasion and exist, if at all, as herbarium material.

Those species described in the text are, in the section lists, marked with an asterisk (*). The decision as to whether to describe a species or not has largely been governed by the frequency of reference to the species in the pages of the *Bulletin*. The species which are in general cultivation and those which are on the fringe of cultivation but which might be more readily available with an upsurge of interest in the genus, are described more fully. Species which have been in cultivation somewhat transiently and species which have otherwise created interest, such as *P. agleniana*, that 'truly magnificent primula', have received less full treatment.

But not all species referred to in the *Bulletin* have been described, for some species are only mentioned on a single occasion, as when seen fleetingly by a traveller in the mountains. The decision as to whether to include a species or not is, therefore, a personal one, for which the writer must be held responsible.

The names of the various species of *Primula* have now become fairly well established, and these are the names listed in the preamble to each section. Synonymy has been omitted from the sectional lists, but is included in the general Index. In this connection the unfortunate case of *P. nutans* must be mentioned. This Soldanelloid species has been *P. nutans* for more than ninety years. By one of those whims of botanical fantasy, it should now become *P. flaccida*. But it remains *P. nutans* here. Far too often have changes of this nature been accepted 'because they were in accordance with the International Rules of Botanical Nomenclature'.

The real purpose of this publication is to interest people in primulas. Those species which are in cultivation have therefore been described, a process which has involved the use of a certain amount of botanical detail. This has been kept to a minimum, and some aspects have been omitted entirely: the venation of the leaves, the relative depth of colour between the upper and lower surface of the leaves and, perhaps most important of all, the description of the calyx. The character of the calyx is often of diagnostic value.

This generation of interest in primulas needs further qualification. For the intention is to persuade more people to grow more primulas, to acquire the rarer species and to cultivate, propagate and swop them. It is also important to know the deficiencies of one's microclimate and habitat, both of which can vary tremendously in Britain: and it helps to know something about the natural habitat and climatic conditions under which a species has evolved, an aspect which, together with the story of the discovery and introduction of the primulas would, in itself, fill a volume.

It may therefore be of interest to refer to a very few publications which are relevant to the discovery and introduction of species of the genus *Primula* – and other genera too. In the first place must be mentioned the works of Reginald Farrer and Frank Kingdon-Ward. Both of these travellers produced works too numerous to specify, but most of them are listed in the Society's Library Catalogue. Although there are passages of pure poetry in some of Farrer's works, some find much of his style florid and disappointing. Kingdon-Ward, that intrepid traveller, I think was more of an explorer than a botanist and in his published works spends more time with the country than with the plants. But there are two volumes which I find absolutely fascinating. In the first place there is Cox's *Plant Hunting in China* (8) published in 1945; and *A Quest of Flowers* (9), the plant explorations of Frank Ludlow and George Sherriff – by Dr.

H. R. Fletcher, the same Fletcher of Smith and Fletcher (5), (6) and (7).

One other exclusion. By reason of the Society's title, the species in which we are interested must, in their natural habitat, dwell in alpine or sub-alpine country. Species such as *P. malacoides* – a weed of paddy (rice) fields – are therefore named only.

Finally, the species must all be Asiatic, which still allows of tremendous interest, for the major development of Primulaceae is in western China and, secondarily, in the Himalaya.

This is a book for gardeners. It is assumed that the reader will wish to learn as much as possible about the cultivation of primulas he possesses, and to be assisted in their identification, and also to learn something about primulas whose names he may come across in seed lists, catalogues, etc. The problem in seeking to assist the reader in these latter two aims is to steer a middle course between being too botanical for the gardener-reader and so over-simplifying as to mislead. Therefore, while botanical terms have been kept to a minimum, some are necessary, and these are defined (in non-technical terms), and illustrated, in a short *Glossary* on p. 150.

Keys have been introduced to assist in identification. These are simple, both in construction and in content. In fact, they contain little or nothing that is not in the text, so that the reader who does not refer to them is losing nothing except the convenience of being able to identify his plant without blindly hunting through the text. They are based upon leaves and flowers (not the roots, fruit, or anything requiring a microscope or lens): this narrow selection of identifying characters must mean some loss of precision, and this, together with the fact that a good many intermediate forms have arisen from seed sown in this country, is bound to mean that at times a plant will not 'key out' correctly.

The *Key to the Sections* (logically the first requirement) has presented special difficulty, because the differences between sections depend to a considerable extent upon *combinations* of characters, and also upon factors not readily assessable by the non-botanical gardener – the manner in which the leaves unfurl, detail of fruit and seed, the microscopic characters of the chromosomes, etc. Nevertheless such a key has been attempted (p. 37), and it is in fact that of Smith, W. W. and Forrest, G., which was published in *Notes on the Royal Botanic Garden, Edinburgh*, Vol. 16, in 1928. This key has been slightly modified to take account of revisions of the sections since that time.

[4]

It is to be emphasised that both the keys and the glossary are designed to apply only to the species of primula described in this book, except that in the key to the sections all sections are included. With the *Index* it is quite otherwise: the aim here is to provide a reference list to all the names which have been used (as specific epithets – not including sub-species, varieties and forms) for Asiatic primulas: for this purpose it is as complete as the writer can make it. This has been done in the belief that such a nominal list will meet a need for reference purposes.

It is necessary to stress the limited *geographical area* covered by the book. Asia itself does not seem to be a very clearly delineated continent so far as its boundary with Europe is concerned. The boundary seems to lie in part in the Caucasus Mountains, and plants from that area are not included in this book – nor indeed are many of the West Asian species, whose affinities tend to lie with the Caucasus primulas. It is hoped that all these will be included, together with the European and American primulas, in the companion book which the Society intends to publish in the near future.

CULTIVATION

Many, but by no means all, of the Asiatic primulas are difficult plants to grow. Fortunately, the species of the various sections tend to occupy a rather similar natural habitat, so that cultural requirements can usually be treated on a sectional basis rather than on a specific basis. Where the difficulties or requirements of the species of a section demand special attention, this has therefore been commented on under the particular sectional heading.

Before proceeding to discuss the cultivation of primulas, a word about growth and form may prove helpful. The stature given for the various species is derived from the standard works of Smith & Fletcher (5), (6) and (7). The dimensions are usually based on an examination of many plants of species growing in the wild. In cultivation, a kindlier habitat may result in a more vigorous plant and, equally, unsuitable growing conditions may produce a dwarf, stunted form.

In the wild, altitudinal and geographical factors tend to keep the species true and there is comparatively little hybridization. In cultivation there are, of course, no such natural barriers and hybridization is much more frequent, but almost always between species of the same section, and not between sections. Vigour, form and flower colour of the hybrids are intermediate between the same characters of the parent species.

With some species, particularly the more widely distributed species, various more or less consistent forms of the type have developed and some of them have been given subspecific or varietal status. Details of these are given in the standard works. Extremes of altitudinal range and geographical isolation are the two main factors resulting in the development of these forms.

There are two main periods of hazard in the cultivation of Asiatic primulas: the dormant, winter period and the period of growth and flower from spring to autumn. The Asiatic primula is therefore at risk the whole year round, but the hazard differs with the season.

WINTER

In the natural habitat, most Asiatic primulas are covered for from four to six months of the year with a thick blanket of snow. Under these conditions they have a long, cool resting period, but are not hard frozen, and they are damp but not wet. In the north of England, winter is characterized by periods of cold grey wetness and by periods of frost. In other parts of Britain there are prolonged periods of hard frost with no snow cover, and then again there may be periods of excessive dryness – a condition more usually associated with summer, when it is perhaps even more of a hazard.

The success with which a primula resists our British winters depends upon the inherent stamina of the plant and on its form of winter growth, which may vary within the same section. These forms comprise the completely herbaceous species, species persisting as a winter-resting bud, species with a winter-persistent, but much reduced, winter-green rosette of leaves and species with completely different, much contracted and often farinose leaves.

The species most easily maintained are those which are completely deciduous. In late autumn, a little bonemeal may be pricked into the soil and a light mulch of leafmould or peat applied – and the position of the plants should be marked. But not all deciduous species have the necessary stamina. The species of the section Soldanelloideae (on which a special note is included) are a case in point – except *P. nutans*, which has much greater vigour.

Species persisting as a resting bud, of which the Nivalids are the best known example, are much more difficult. After the leaves have withered in autumn, the bud should be pushed well into the ground, and ringed and covered with dry peat. A frame light, cloche or sheet of glass can also be used to give protection. But with *P. sonchifolia*, a Petiolarid which has much tougher, clasping scale leaves, the problem is not protection of the bud, but rather the fact that the anchoring roots also seem to diminish in this winter-dormant state and the buds are scratched out by the birds.

Species which persist as a contracted winter-green rosette vary considerably in their resistance to our winters. Species such as *Pp. ianthina* and *helodoxa* (both Candelabra) and *P. secundiflora* (Sikkimensis), which have a reasonably vigorous rosette of leaves, are perfectly easy. It should, however, be noted that most primulas of the Sections Sikkimensis and Candelabra are herbaceous.

Many species of the section Farinosae persist as quite small, tight rosettes which are liable to be lifted by frost. They have to be pushed back into the ground and as they have no great stamina they ought to be given some protection with peat. It is also wise to take seed of the smaller species of this section because, even with care, many are short-lived. These small species can also be grown in pots, which gives added winter protection.

Then there are species such as the Capitatae, which in autumn appear to be growing quite happily as multi-crown plants of dwarf stature. When the leaves contract to the winter form, the rosettes are seen to be at the end of horizontal, more or less twiggy stalks which radiate like the spokes of a wheel from the central rootstock. These rosettes can be at risk. The rootstock should be carefully pushed into the ground and the rosettes and stalks protected with peat.

Finally, there are the species which in summer have a more or less green leaf of moderate vigour, but in winter a much contracted leaf usually of quite different form and often heavily silver- or yellow-farinose. These species are typically the Petiolarids, on which a separate note has been given. These plants can be grown in pots, but they grow better outside, even in winter. They should be planted in a face as nearly vertical as possible and given some overhead protection, which might only be a peat-block or a piece of flagstone; but the flowers are very early, and very beautiful, and they can get rather battered; and they are also more or less sessile and therefore difficult to see and perhaps cannot be fully appreciated. In the cool alpine house there is fully controlled protection against weather, and the plants can be lifted and the flowers enjoyed. But water on the foliage spoils the farina and rots the leaves.

I think I have the answer. I was recently obliged to go abroad in midwinter, and the Petiolarids in the alpine house got no water at all from mid January to mid February. All the foliage of both farinose and efarinose species was perfectly healthy, with no sign of decay at all, and *P. edgeworthii alba* flowered profusely from mid January to mid March. My Petiolarids, during the winter, are in clay pots plunged in about 18 in. of sand, which retains some capillary moisture. This is surely a case for bottom-watering, and not too much of it.

Spring to Autumn

This, of course, is the season of growth and flower and of the production of seed. Many species are quite easy to cultivate, but the majority probably require that little extra care and attention to produce a really good plant, and some species are almost impossible. The key to success is undoubtedly an understanding of the natural habitat and of the differences in the habitat one is obliged to provide. Fortunately, in this matter also, the species can be treated sectionally, for the members of the various sections affect similar growing conditions, or there may be two or three areas of preference within some of the larger sections.

The chief difficulty in growing primulas during the summer months lies in regulating the incidence and amount of moisture and, with some species, the manner in which moisture is made available. Most of the Asiatic primulas with which we are concerned are high alpines of the mountains. In the early weeks of growth the plants are fed by the copious melt-water of the snows. The plants are well supplied with cold ground-water, but there is no precipitation. Most of the primulas flower in this period. And then comes the Monsoon.

The Monsoon is a strong south-westerly air-stream which, heavily charged with moisture, sweeps up the Indian Ocean and across the Bay of Bengal. The chief impact is on the main Himalayan Range, and some of the hills of Assam, which the Monsoon strikes before reaching the Himalaya, have a rainfall of more than 300 inches during the months of June to August.

The major precipitation is against the central and eastern Himalaya. Once the height of land is passed the climate becomes drier and cooler, and much of the plateau of central Tibet is cold, dry desert. But not all the rainfall is stopped by the Himalaya. Across the eastern end of the range some 60–80 inches of precipitation travels into the great river-valleys of western China.

On the plains of northern India, both the start and the finish of the Monsoon is quite sudden and there is no rainfall at all at other times of the year. But in the mountains, the period is less well defined: the rains are heralded by periods of storm and die away in days of sunshine, rain and mist. There are other variations, too, in the pattern of climate. Kashmir and the western Himalaya, for instance, lie outside the main area of Monsoon influence and summer rainfall consists of periods of storm, often with long periods of drought between. Primulas from these areas are therefore more drought-resistant.

The primulas with greatest thirst are species of the section Sikkimensis, followed closely by Candelabra. If these species are grouped by the side of a stream or on the edge of a pond, where the roots can get down to water, then magnificent plants can be grown, even in an open, fairly sunny position. But these facilities usually do not exist in the small garden and the only answer is to grow in shade, with the ground well watered and mulched.

[9]

Another difficulty, of an entirely different nature, arises with some of the primulas of lesser stamina, perhaps particularly with species of the sections Farinosae and Soldanelloideae. These dwarf plants are sometimes described as short-lived perennials, but they tend to behave as annuals or biennials. The plant seems to realise that it has little chance of survival and so puts everything into the production of viable seed. If this seed is collected, sown and successfully raised, then the species persists in cultivation. But all too often something has gone wrong with the cycle, and a species has been lost. If fresh wild seed can be imported, then the default can be remedied, but access to the sources of wild seed is, politically, particularly difficult at the present time.

In addition to the difficulties described above, which are peculiar to the genus, there should also be adequate protection against the usual run of garden pests including wood-lice, wireworm, cockchafer grubs, sawflies and other leaf-cutting predators, and slugs – many of the lesser, moisture-loving species seem to have a particular attraction for slugs. The usual remedies apply, defence having been considerably increased by the availability of systemic insecticides such as the various products containing malathion.

There is another pest which certainly used to be rather serious – primula root aphis, an attack of which seriously weakens and probably kills an infected plant. The aphids appear to be somewhat migratory between the roots and the neck of the plant. Mrs. C. B. Saunders (10) advocates the use of paradichlorbenzine, more easily referred to as PDB and most readily available as Jertox moth crystals. This product is a white waxy substance which can be split between the thumb and forefinger and subsequently powdered. It is pricked into the soil surface around infected plants, and around clean plants as a preventive – I have used it for this latter purpose but, fortunately, have no personal experience of the root aphis. PDB, which has a powerful chemical smell, sublimates into a heavy vapour which penetrates the surface layers of the soil. It has been suggested that the systemic malathion preparations can also be used to control primula root aphis.

Another product, only recently available, is Gesal Root Guard, a Ciba-Geigy product which reputedly 'kills all common below-ground pests' and has an estimated life of 8–10 weeks. This would appear to be a very useful addition to the armoury, but I have used it too recently to be able to assess its possibilities.

PROPAGATION

Most, but not all, primulas are fairly easily grown from seed. For the compost a good John Innes No. 1 may be used. Some discrimination should be exercised in selecting the compost, as the product varies. It should be friable, and contain no clay.

My own personal experience is that clay pots give much better germination than do plastic pots. Moist compost is placed in the pot, and levelled and firmed using the base of another pot. Seed is broadcast thinly on this prepared surface, given a very thin covering of compost and firmed again. The pot is then given a covering of granite chips, which helps to conserve surface moisture and also prevents the growth of moss, which might otherwise choke the seedlings, on the surface of the soil; although a disadvantage is that surface dryness cannot readily be seen. The granite chips have to be scraped away and the finger-end pressed into the soil, or the pots tapped, or a Diplex type soil-moisture meter used.

Another way of sowing primula seed, particularly fine seed, is to firm the compost as before and then place a thin layer of granite chips over the surface of the compost. The seed is broadcast over the stone chips, which must not be hard-packed, and contact with the compost is obtained by fine-watering from above.

The pots should be kept in a cool, shaded place, but not necessarily in total darkness. Beneath the staging in the alpine house is quite suitable, provided the pots are not dripped on. The pots are also usefully sunk in sand to reduce drying out. The area beneath the staging may be blocked off with a dwarf wall and filled with sand for this purpose.

The pots should be kept damp but not wet, and when seedlings begin to appear protection should be given against such pests as slugs and woodlice. With respect to slugs, in addition to the long used poisons based on metaldehyde, there is a recently developed and more persistent product called methiocarb ('Draza').

The seedlings should be pricked out when the first pair of true leaves are developed. The first pair of leaves are the seed leaves. These are of simple profile, and soon aborted. The seedlings should be handled by the leaves, and not by the roots. They should be kept cool, damp and ventilated.

It is sometimes suggested that primula seed should be frozen to assist germination and indeed in the United States this appears to have been developed to a fine art, with temperature and period of freezing for optimum germination of the various species being determined. But at the Royal Botanic Garden, Edinburgh, the freezing procedure is not used. On the other hand, the successful germination of seed is in no way impaired if the seed gets frozen naturally for the period for which frosts usually last in this country.

With respect to the time of sowing, better germination comes from an autumn sowing, but most of the seedlings are herbaceous and have difficulty in building up a sufficiently robust root-stock to carry them through the winter and break away into growth in spring. Seeds sown in early spring have a reduced percentage germination, but all seedlings obtained should grow away into sturdy little plants before the period of dormancy is reached.

While the above remarks as to the ease of growing primulas from seed apply to most species, some of the more intractable species have been raised as seedlings, but made no further progress. There are also certain well-known exceptions to the rule of growing primulas from seed. The Petiolarids are the best-known example. Species of this section flower very early in the year. The capsule disintegrates and the seed is washed out, hopefully to germinate very quickly and build up sufficient stamina before the onset of the rains. The lesson is that the seed of the Petiolarid primulas should be sown green. Viability deteriorates rapidly with age.

With respect to the production of seed, many species of primula do not set seed very readily. Primulas are not self sterile, but rather are they either protandrous or protogynous, i.e. either the stamens mature before the ovary or the ovary matures before the stamens. A single plant therefore rarely sets fertile seed and it is advisable to hand pollinate rare species. On the other hand some species hybridize fairly readily, particularly species of the same section which flower at the same time. In these circumstances it may be necessary to hand pollinate and bag the flower scape if it is required to keep species true from seed. A fertile ovary swells considerably to form a tight, fat capsule of spherical or sub-cylindrical form, and if this swelling is not observed then seed formation within the capsule is abortive.

In the text mention has been made of the fact that the various species are either dimorphic, having both 'pin-eyed' and 'thrum-eyed' forms, or monomorphic, having one form only. This feature concerns the relative positions of anthers and stigma (see illustration

p. 152), and it is significant in relation to self-fertilisation. The chances of self-fertilisation (or of fertilisation of one flower by another on a single plant) in dimorphic species are said to be in the order of 5%, as against something like 100% in favourable conditions when a pin-eyed (long-styled) flower is crossed with a thrum-eyed (short-styled).

Vegetative Propagation

When plants of any one species are grown from seed, they all vary slightly, but within the overall characteristics of the species. Sometimes variation lies in the colour of the flower: there may be good colour forms and there may be poor colour forms. There may be variation in habit of growth: some forms are dwarf and compact, other forms are loose and gross. Or again there may be variation in the vigour of the plants produced from any one batch of seed: some may be miffy and others good doers. Whatever the reason, one may wish to select one particular plant out of a batch of seedlings, to propagate that particular form and dispose of the rest. In such circumstances, some form of vegetative propagation has to be employed.

Many species can be divided, or carefully cut and pulled apart. With the Petiolarids, which flower very early, this should be done after the flowering period is over. With the later-flowering species one usually has a choice: a plant may be divided before or after flowering. If division is carried out when growth first appears in spring, then the individual crowns are more easily separated and there is not the mass of foliage to keep going as is the case with later divisions. Depending on the vigour of the species, the divided fragments may be planted in the open ground or potted and kept in the alpine house in a cool, shady corner, or even in a closed case. It is not only the vigorous species which can be treated in this manner. I have successfully divided plants of *P. reidii* var. *williamsii* on several occasions. Division of a plant before flowering usually delays flowering, and may result in loss of flowers for the one season.

Alternatively, plants may be divided after flowering, when the foliage is green and healthy. Some of the grosser leaves should be stripped off the divisions and, again depending on the vigour of the species, the divisions may be planted in the open garden or potted and kept in the alpine house. At this time of year much more attention has to be paid to watering, and shade always helps. Some of the more vigorous species, *P. denticulata* for example, actually benefit from being broken up and propagated in this way, every three or four

[13]

years. Only the outer, more vigorous and youngest crowns should be retained, and the central rootstock thrown away.

Then there are the primulas which overwinter as a group of big, resting buds. *P. sonchifolia* is the best example. It is so easy to divide the buds and re-plant or re-pot. But in this resting condition the buds have a tenuous root system of fine white hair-like growths. If this root-system is torn away from the buds, as is quite easily done, then the flowers may begin to open and the plant suddenly finds it cannot supply sufficient food. Extra care with additional watering and some shading may therefore be necessary with these divisions.

Most primulas can also be propagated by means of root cuttings. For this purpose a plant, or part of a plant, is dug up and the soil washed off the roots. The more or less thong-like roots are then cut into sections the length of which depends on the vigour of the plant. The top of each section is cut square, and the bottom oblique, so that the top and bottom of the cuttings can be identified. The standard procedure is to place the cuttings in a pan of sand and peat with the cuttings upright and the top of the cutting level with the top of the sand. The pan has to be kept close for a few weeks, and air subsequently admitted. But a better way is to put the cuttings in a polythene bag, with damp chopped sphagnum. With this method you can see the formation of the root and leaves. The cuttings are subsequently taken from the polythene bag, anchored in a leafy compost and kept cool and moist.

There are also various specialist methods. Masterton describes the propagation of petiolarid primulas by leaf-cuttings (11). All that is required is a closed propagation frame – even a box with a sheet of glass over it, but the frame must be shaded. The leaf is held by the base of the petiole, which is pulled downwards. The leaf-bud 'is usually quite small and situated on the medial surface of the petiole at its very base; usually it is lighter in colour than the petiole itself. The leaf-bud must be secured before the leaf will root and develop into a plant.

'The leaf is then inserted in a pan containing sand . . . but not too deeply. With cuttings lined around the edge of the pan, root formation usually takes place within 3 to 6 weeks and the cutting can then be transferred to a leafy sandy mixture.

'Petiolarid primulas treated in this way include *Pp. whitei*, *edgeworthii, scapigera, bracteosa, gracilipes*, × *scapeosa* and *aureata*. *P. sonchifolia* does not appear to produce leaf buds.'

David Livingstone's experiences with *P. bracteosa* might also be

[14]

usefully noted (12). In the fruiting stage, some forms of *P. bracteosa* produce a vegetative bud at the top of the scape which, if pegged to the ground, will root in about three weeks' time, thus providing an easy method of propagation. Plants so propagated are quite as vigorous as those increased by division.

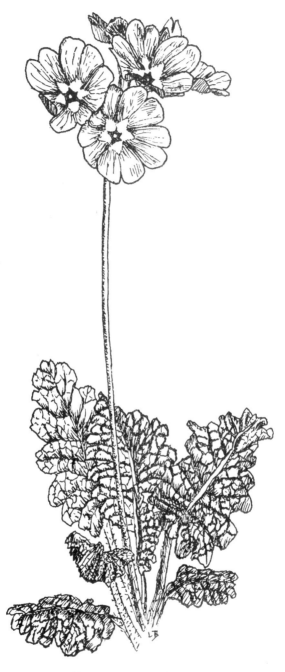

P. sieboldii (Sect. Cortusoides) (× 1) (p. 43)

[16]

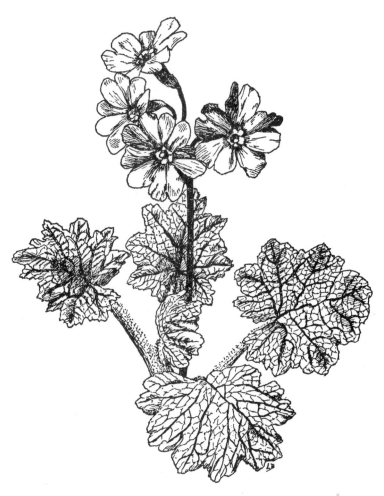

P. polyneura (Sect. Cortusoides) (× 1) (p. 41)

P. reinii (Sect. Reinii) (× 1) (p. 46)

P. edgeworthii (Sect. Petiolares) (× 1) (p. 64)

P. kingii (Sect. Amethystina) (× 1) (p. 53)

P. cuneifolia var. *heterodonta* (Sect. Cuneifolia) (× 1) (p. 54)

P. modesta (Sect. Farinosae) (× 1) (p. 88)

P. primulina (Sect. Minutissimae) (× 1) (p. 78)

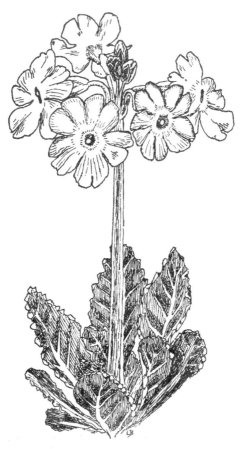

P. gemmifera (Sect. Farinosae) (× 1) (p. 91)

P. denticulata (Sect. Denticulata) ($\times \frac{2}{3}$) (p. 99)

[23]

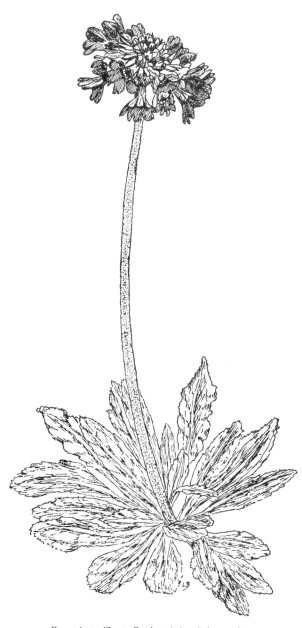

P. capitata (Sect. Capitata) (× 1) (p. 100)

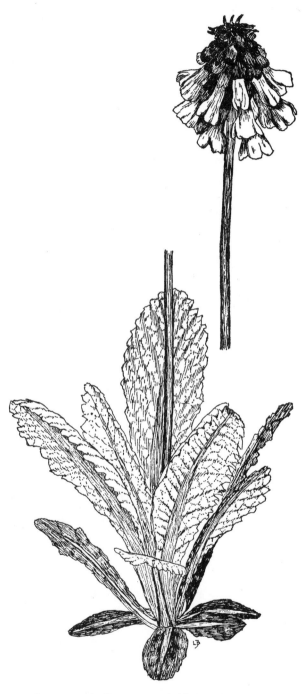

P. muscarioides (Sect. Muscarioides) (×1) (p. 105)

P. nutans (Sect. Soldanelloideae) (× 2/5) (p. 110)

[26]

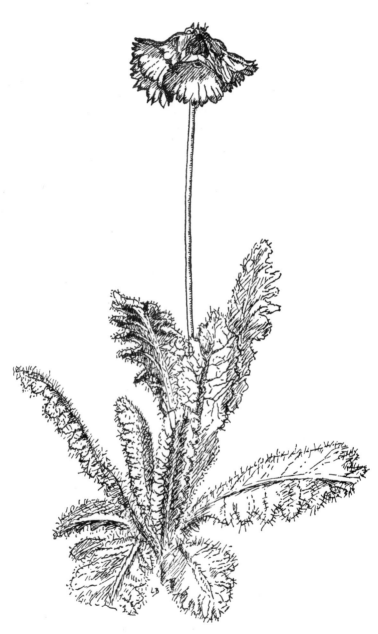

P. wattii (Sect. Soldanelloideae) (× 1)

P. rotundifolia (Sect. Rotundifolia) (× 1) (p. 114)

P. chionantha (Sect. Nivales) ($\times \frac{1}{2}$) (p. 119)

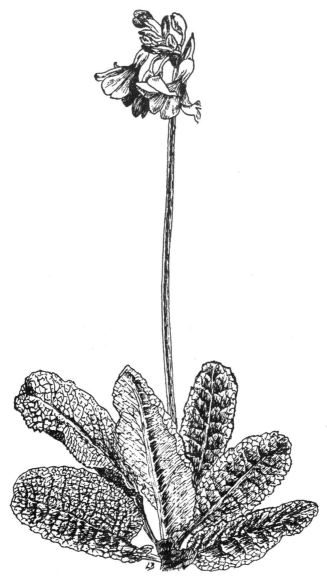

P. alpicola (Sect. Sikkimensis) (×⅔) (p. 130)

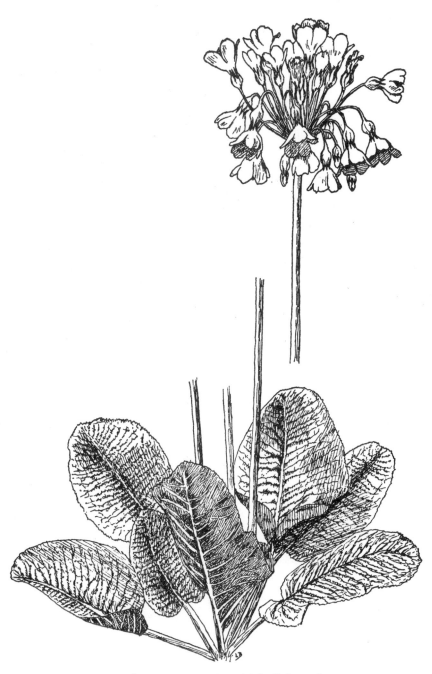

P. florindae (Sect. Sikkimensis) (×½) (p. 131)

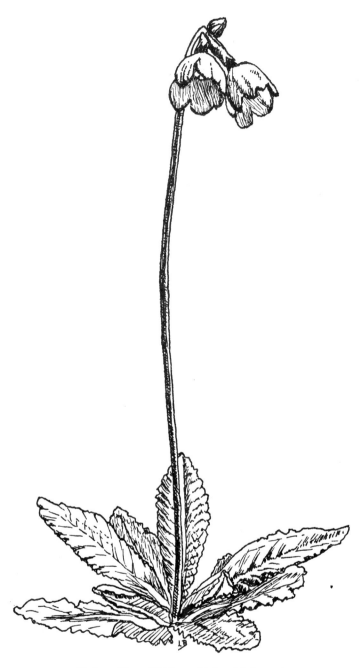

P. ioessa (Sect. Sikkimensis) (×4/5) (p. 132)

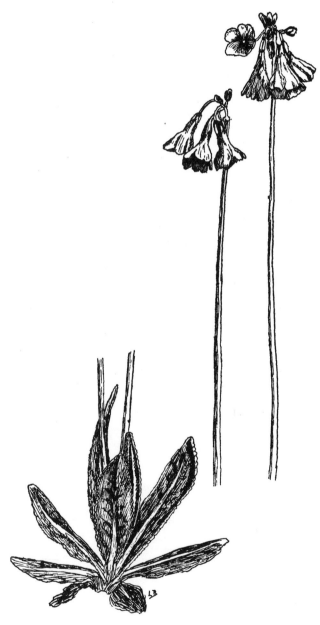

P. secundiflora (Sect. Sikkimensis) (×½) (p. 133)

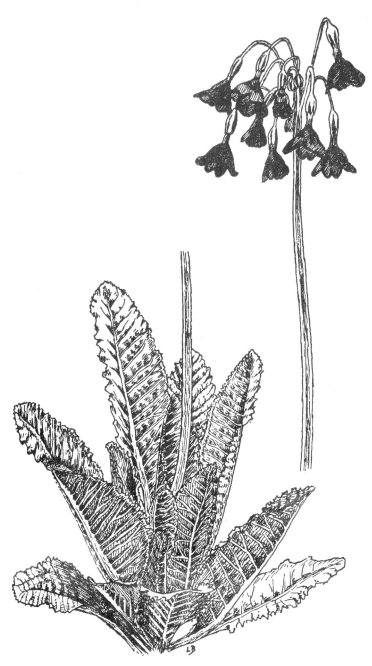

P. waltonii (Sect. Sikkimensis) (×½) (p. 134)

[34]

P. cockburniana (Sect. Candelabra) (×½) (p. 141)

P. pulverulenta (Sect. Candelabra) (×⅔) (p. 144)

[36]

SECTIONS OF THE GENUS PRIMULA

KEY TO THE SECTIONS

Adapted from Smith and Forrest (4)

1 Young leaves folded inwards	2
Young leaves folded outwards	5
2 Flowers in superposed whorls; bracts leafy; corolla-tube much longer than calyx, limb small	Floribundae
Flowers in truss or simple umbel; bracts not leafy	3
3 Corolla usually annulate. Leaves much indented	Cuneifolia
Corolla usually exannulate. Leaves generally more or less entire	4
4 Bracts conspicuous	Parryi (America)
Bracts small	Auricula (Europe)
5 Flowers bell-shaped or cylindrical, with no 'eye'	6
Flowers funnel-shaped with distinct 'eye'	9
6 Flowers cylindrical	Grandis (Caucasus)
Flowers bell-shaped	7
7 Flowers sessile	Soldanelloideae
Flowers with individual stalks	8
8 Inflorescence farinose	Sikkimensis
Inflorescence efarinose	Amethystina
9 All parts efarinose	10
Farina present (sometimes scanty)	18
10 Leaf-blade not distinct from stalk (except *megaseaefolia*)	Vernales (Europe)
Leaves very distinctly stalked	11
11 Leaves more or less lobed	12
Leaves not lobed	Carolinella
12 Leaves pinnate	Pinnatae
Leaves not pinnate	13

13 Calyx longer than corolla	Pycnoloba
Calyx shorter than corolla	14
14 Calyx very wide at base	Sinenses
Calyx narrowed at base	15
15 Calyx cup-shaped, leafy	16
Calyx tubular, rigid	17
16 Flowers in many superposed whorls, or in racemes	Malvacea
Flowers generally in simple umbels	Obconica
17 Seed-capsule more or less enclosed in calyx	Cortusoides
Seed-capsule projecting far out of calyx	Reinii
18 Flowers more or less sessile; bracts without a pouch	19
Flowers with individual stalks; bracts sometimes pouched	23
19 Corolla throat with ball of hairs	Minutissimae (Subsection Bella)
Corolla throat open	20
20 Massive rhizome; inflorescence few-flowered	Dryadifolia
No rhizome; inflorescence many-flowered	21
21 Flowers in dense spikes	Muscarioides
Flowers in dense rounded heads	22
22 Flowers more or less erect	Denticulata
Flowers more or less pendant	Capitatae
23 Seed-capsule round, crumbling, not opening by teeth	Petiolares
Seed-capsule round or cylindrical, opening by teeth	24
24 Massive woody rhizome; leaves more or less bullate; bracts leafy	Bullatae
No woody rhizome	25
25 Leaves more or less lobed or strongly indented, stalked	26
Leaves not lobed or deeply cut	27
26 Seed-capsule round, enclosed in calyx	Malacoides
Seed-capsule cylindrical, projecting from calyx	Souliei
27 Leaf-blade rounded, heart-shaped at base, petiole long and distinct	Rotundifolia
Leaf-blade not rounded, tapering into stalk	28

[38]

28 Bracts generally more or less pouched	Farinosae	
Bracts never pouched	29	
29 Minute species; often solitary flowers	Minutissimae (Subsection Eu-minutissimae)	
Robuster species; inflorescence many-flowered	30	
30 Seed-capsule round	Candelabra	
Seed-capsule cylindrical	Nivales	

1 THE SECTION CORTUSOIDES

THE SUBSECTIONS AND SPECIES

SUBSECTION EU-CORTUSOIDES

P. cinerascens	P. lactiflora	*P. saxatilis
*P. cortusoides	*P. mollis	*P. sieboldii
P. eugeniae	*P. polyneura	P. violaris
P. kaufmanniana		

SUBSECTION GERANIOIDES

P. alsophila	*P. kisoana	P. palmata
*P. eucyclia	P. latisecta	P. pauliana
*P. geraniifolia	P. loeseneri	P. septemloba
*P. heucherifolia	P. minkwitziae	P. vaginata
*P. jesoana	P. normaniana	

* In this and subsequent Sections the asterisks indicate the species described.

KEY

1 Leaf veins pinnate (Eu-Cortusoides)	2	
Leaf veins basal (Geranioides)	6	
2 Leaves as broad as long, or nearly so	3	
Leaves distinctly longer than broad	4	
3 Leaves deeply lobed	polyneura	
Leaves scarcely lobed	mollis	
4 Calyx lobes erect	5	
Calyx lobes spreading	sieboldii	
5 Pedicels 1 cm. or less at flowering time	cortusoides	
Pedicels more than 1 cm. at flowering time	saxatilis	
6 Flowers more or less annulate	7	
Flowers exannulate	10	

7	Plant hairy. Petioles not sheathing stem	8
	Plant almost hairless. Petioles sheathing stem	*eucyclia*
8	Corolla hairy	*kisoana*
	Corolla not hairy	9
9	Flowers usually in a single umbel	*geraniifolia*
	Flowers usually in 2 or more umbels	*jesoana*
10	Leaf lobes and teeth rounded or obtuse	*heucherifolia*
	Leaf lobes and teeth more or less acute	*geraniifolia*

With the exception of the type species *P. cortusoides*, which ranges from Siberia in the east to west of the Urals in European Russia, the distribution of species of this section is entirely Asiatic and extends from Japan and Korea, through Siberia and Central Asia to the northern and western provinces of China, Tibet and the eastern Himalaya.

The rootstock of species of this section is a usually short, slender to stout rhizome. The efarinose leaves are green. The blade of the leaf is more or less round or orbicular, with margin variously lobed. The long slender leaf-stalk or petiole equals or is longer than the blade, and the foliage is softly and shortly hairy. The flowers are carried in one, or occasionally two or three superposed, umbels of flowers at the head of a scape.

Many species of this section are dwarf woodland plants, which suggests some degree of moist shade in cultivation. Other species are of coarser growth, with more deeply indented margins and heavier venation of the leaves.

All the species of this section are entirely herbaceous, and the dwarfer species can be usually interplanted, for example with such subjects as *Cyclamen coum* and *Lithophragma parviflora*.

This section has been divided into two subsections, according to the arrangement of the veins of the leaf and the nerves of the calyx. Species with the lateral veins of the leaves pinnate belong to the subsection Eu-Cortusoides, whereas species with lateral veins mostly basal belong to the subsection Geranioides. The distinction is a minor one, but useful in grouping the species. The section Cortusoides is separated from adjacent sections by the structure of the calyx.

THE SUBSECTION EU-CORTUSOIDES

P. cortusoides Linn.

This is the type species of the section. The foliage is rather similar to that of *Cortusa matthioli* (which belongs to the primula family but is not a primula) – whence the origin of the specific epithet.

P. cortusoides is a more or less hairy perennial. The leaves, including the petiole, are 6–26 cm. long and 1.5–6 cm. broad. The long thin petiole is about twice the length of the lamina. The blade is ovate to oblong, cordate at the base and rounded or obtuse at the apex. The margin is crenately lobed, the lobes being irregularly toothed.

The scape is 10–15 cm. tall and carries an umbel of 2–15 flowers. The pedicels are 5–10 mm. long, lengthening to 2.5 cm. in fruit. The umbel of *P. cortusoides* is therefore a fairly tight head of flowers.

The annulate corolla is rose or red to rose-violet, and pin-eyed and thrum-eyed forms occur. The corolla-tube is 1–1.5 cm. long and the limb 1.5–2 cm. diameter with obcordate lobes which are notched at the apex. The calyx-lobes are erect.

This is one of the dwarf, woodland plants – neat ground cover. It was first brought into cultivation in Britain in 1797 from Siberian seed sent back by Professor Pallas.

In cultivation, *P. cortusoides* is often confused with *P. polyneura* and *P. saxatilis*.

P. mollis Nutt. ex Hook.

This species was discovered by Booth, in about 1852–3, in the mountains of Bhutan where it grew, at an elevation of 2300–3300 m., in shady places at the edge of thickets and by the larger streams.

Seed was sent back to Nuttall, who raised plants from which the species was subsequently established in cultivation: it was not seen in the wild again until 1905.

This is a very hairy perennial with rather coarse foliage. The scape is 10–60 cm. tall, there being as many as 6 to each plant. Each scape carries 2–10 superposed umbels, each umbel having 4–9 flowers. The pedicels are 1–3 cm. long.

The annulate corolla is dark rose to deep crimson with a yellow or greyish-green eye. The corolla-tube is 1–1.5 cm. long and the limb 1–2 cm. diameter with obcordate lobes which are notched at the apex.

This species is doubtfully hardy even in the milder parts of Britain, but it grows vigorously under glass and there sets seed.

P. polyneura Franch. (Illustration p. 17)

This species was described by Franchet in 1895, in part from material collected by Soulié near Tatsienlu and in part from material collected by Prince Henri d'Orléans.

P. polyneura is a densely hairy to almost glabrous perennial, slightly more robust than *P. cortusoides*. The leaves, including the petiole, are 4–30 cm. long and 2–10 cm. broad. The petiole equals or is one and a half times the length of the blade. The lamina is broadly ovate or orbicular, cordate at the base and 7–11 lobed, the lobes being broadly ovate, with margin more or less entire or conspicuously toothed.

The slender to stout scape is 10–50 cm. tall and carries 1–3 umbels, each of 2–12 flowers. The pedicels are 5–25 mm. long.

The annulate corolla is pale rose to crimson and wine-purple. The corolla-tube is 1–2 cm. long and the limb is 1–2.5 cm. diameter, with obcordate lobes shallowly notched at the apex and sometimes shallowly toothed on the margin.

The species is fairly common in the Provinces of Kansu, Szechuan and Yunnan in Western China.

P. saxatilis Komarov

P. saxatilis was described from plants collected by Komarov in 1897. These were growing in rock-fissures rich in humus, on cliffs of the basaltic plateau of Pei-shan in North Korea. There is also some evidence of the occurrence of *P. saxatilis* in some of the northern provinces of China.

This species is a more or less hairy perennial. The leaf, including the petiole, is 6–20 cm. long and 3–6 cm. broad. The petiole is rather less than twice the length of the leaf-blade. The lamina is oblong to ovate-oblong, cordate at the base and lobed at the margin, the lobes being entire to irregularly toothed.

The stout scapes are 10–30 cm. tall, 1–6 to each plant. Each scape carries an umbel, or sometimes two superposed umbels, of 3–15 flowers. The slender pedicels are 2–4 cm. long and more or less erect, or drooping.

The annulate corolla is rosy-violet to pinkish-mauve. The corolla tube is about 1 cm. long and the limb 1.5–2.5 cm. diameter, with obcordate lobes deeply notched at the apex. Pin-eyed and thrum-eyed forms occur. The calyx lobes are erect.

Smith and Fletcher (6) say that *P. saxatilis* '. . . is well known in gardens, divides readily and ripens seed freely'. But it is not now widely offered; perhaps this is because there is a certain coarseness of foliage, and the colour of the flower can be unfortunate; but it is always worthwhile flowering a batch of seedlings.

P. sieboldii E. Morren (Illustration p. 16)

The distribution of this species is Japan, Korea, the basin of the river Amur, Manchuria and Northern Mongolia.

P. sieboldii is a rather hairy perennial. The leaves, including the petiole, are 4–20 cm. long and 2–7 cm. broad. The petiole is equal to, or a little longer than, the leaf-blade. The lamina is ovate to oblong-ovate, cordate at the base and rounded at the apex, and the margin regularly lobed, the lobes being irregularly toothed or shallowly lobed.

The stout scape is 6–30 cm. tall and carries an umbel, or sometimes two superposed umbels, of 2–15 flowers. The pedicels are 5–40 mm. long and more or less erect.

The annulate corolla is rose-violet to lilac-purple, or deep crimson with a white eye, or sometimes pure white. The corolla-tube is about 1 cm. long and the limb 1–3.5 cm. diameter, with broadly obcordate lobes deeply notched at the apex and margin entire or shallowly lobed. Pin-eyed and thrum-eyed forms occur. The calyx lobes are spreading.

The species was introduced in 1862 by the Chevalier P. F. von Siebold who cultivated it in his garden at Leyden, but the material was subsequently taken over by the London firm of Veitch. In the latter part of the nineteenth century the plant received a great deal of attention from the horticulturalists, and many varieties and colour-forms were grown; but the species does not now hold its former prominence in Britain.

With me *P. sieboldii* remains something of a question mark for I cannot grow it. I suspect a combination of summer drought and slug attack in spring and autumn. But I know of gardens within a very few miles where *P. sieboldii* freely seeds itself.

THE SUBSECTION GERANIOIDES

P. eucyclia W.W. Sm. et Forrest

Specimens of this species were first collected by Forrest, in 1921, in the Tibetan Province of Tsarong. Of *P. eucyclia* Forrest says 'it grows in crowded colonies, forming mats many yards in extent sheeted with bloom on stony alpine meadows, and is also seen on cliffs'.

Although Forrest obtained seed, no plants were subsequently raised. The plant was again found in 1926 by Kingdon-Ward, in north western Burma, and from his seed flowering plants were raised – one gaining an Award of Merit in 1930. But it is doubtful whether any plants of the species still remain in cultivation.

This is a small plant, the leaves including the petiole being only 2–8 cm. long. The slender scape is 3–6 cm. tall and carries one, or usually two, flowers. The annulate corolla is shell-pink to deep mauve with a yellow or reddish eye.

P. geraniifolia Hook.

This species is a quite hairy perennial. The leaves, including the petiole, are 5–20 cm. long and 3–8 cm. broad. The petiole is about one and half times the length of the leaf-blade. The lamina is orbicular, cordate at the base, and palmately and sharply 7–9 lobed, the lobes more or less irregularly and sharply toothed.

The slender scape is 10–30 cm. tall and carries an umbel, or rarely two superposed umbels, of 2–12 semi-pendent flowers. The pedicels are 5–12 mm. long.

The corolla is rose to purple, with or without an annulus. The corolla-tube is 1–2 cm. long and the limb 1–2 cm. diameter, with obcordate lobes more or less notched at the apex. Pin-eyed and thrum-eyed forms occur.

The species is hardy in cultivation, but not widely offered. This is possibly because *P. geraniifolia* can be a rather coarse plant.

P. geraniifolia is found growing among shady boulders and in forests at an elevation of 3,000–4,000 m. Its distribution extends from Nepal and Sikkim to Bhutan, south-eastern Tibet and Yunnan.

P. heucherifolia Franch.

This species was first collected in 1869 by the Abbé David in Western Szechuan, and appears to be confined to that Chinese Province. It grows at 2000-3200 m. 'in grassy places, in the shade of rocks and occasionally in woods' (6) and therefore appears to have a preference for shade.

P. heucherifolia is a hairy perennial. The leaves, including the petiole, are 6–15 cm. long and 3–6 cm. broad. The petiole is equal to, or a little longer than, the leaf-blade. The lamina is orbicular, cordate at the base, palmately 7–11 lobed, the lobes ovate-triangular and irregularly and bluntly toothed. The slender scape is 15–30 cm. tall and carries a single umbel of 3–10 nodding flowers. The pedicels are 1–3 cm. long.

The corolla is mauve-pink to deep rich purple. There is no annulus, but there are variously coloured concentric rings at the throat. The corolla-tube is about 1–1.5 cm. long and the limb 1–2.5 cm. diameter, with obcordate lobes shallowly notched at the apex. Pin-eyed and thrum-eyed forms occur.

P. heucherifolia is another species which is not now widely offered.

P. jesoana Miq.

This species was first described in 1867 by Miquel. Its distribution is given by Takeda as the 'mountainous districts of Central Japan'.

P. jesoana is a more or less hairy perennial. The slender petiole is 7–30 cm. long, about three times the length of the leaf blade. The lamina is orbicular, usually deeply cordate at the base and palmately 7–9 lobed, the lobes being shallowly lobed and more or less irregularly toothed.

The slender scape is 1–60 cm. tall and carries 1–4 umbels, each of which has 2–6 flowers. The pedicels are 3–10 mm. in flower, but may be three times this length in fruit.

The slightly annulate corolla is rose or rose-purple with a yellow eye, or sometimes white. The corolla-tube is 1–2 cm. long and the limb 1–2 cm. diameter, with obcordate lobes shallowly notched at the apex. Pin-eyed and thrum-eyed forms occur.

A species that is now rarely offered.

P. kisoana Miq.

Takeda describes this species as 'one of the most beautiful, rare and interesting of the Japanese primulas'. Its distribution is confined to south-western Japan and it is quite a rare plant, even in its native habitat.

P. kisoana is a very hairy perennial. 3–7 leaves grow in the form of a tuft. The length of the leaf, including the petiole, is 5–15 cm. and the leaf stalk is about equal to the blade. The lamina is orbicular, deeply cordate at the base and shallowly lobed, the lobes being more or less crenate, or merely wavy, at the margin.

The scape is 5–20 cm. tall and carries an umbel, or sometimes two superposed umbels, of 2–6 flowers.

The annulate corolla is hairy, deep rose or rose-mauve, and red in the tube. The corolla-tube is 1–2 cm. long and the limb 2.5–3 cm. diameter, with obcordate-oblong lobes deeply notched at the apex. Pin-eyed and thrum-eyed forms occur.

P. kisoana has from time to time been in cultivation in Britain, but always transiently. Cultivated plants do not appear to set seed, and propagation was always uncertain.

It is doubtful if the species still remains in cultivation.

2 THE SECTION REINII

THE SPECIES OF THE SECTION

P. hidakana	*P. reinii	
P. kamuiana	*P. takedana	*P. tosaensis

KEY

1 Flower exannulate, white	takedana
Flower annulate, rose to purple	2
2 Corolla tube 3 × length of calyx; petals notched to half their length	reinii
Corolla tube 1–2 × length of calyx; petals notched to less than ¼ their length	tosaensis

The species of this section form a small but distinct group with petiolate rounded leaves, small calyces and long cylindrical capsules which protrude far beyond the calyx. These species bear a general resemblance to certain species of the section Cortusoides, perhaps more specifically the sub-section Eu-Cortusoides, to which the species of the section Reinii are obviously related. But there are also certain cytological differences which point to the separation of the two sections.

The distribution of all species of the section Reinii is confined to Japan, where they grow in rocky places and mountain meadows on the upper slopes of the higher mountains. But even in the wild, these are fairly rare plants.

P. reinii Franch. et Savatier (Illustration p. 18. Plate III)

This, the type species of the section, is a small, hairy perennial with a short stout rootstalk. In the winter-time, the vegetative parts of the plants reduce to a resting bud protected by a few brown scale-leaves.

The leaves including the petiole are 1.5–26 cm. long and 0.5–7cm. broad, and the petiole is three to four times the length of the lamina. The lamina is orbicular to reniform, widely cordate at the base and shallowly 7–9 lobed, or more deeply lobed; margin crenate to sharply dentate.

The slender scape is 2–10 cm. tall and usually carries a single

[46]

large erect flower, or an umbel of 2–3 flowers. The pedicels are 1–3.5 cm. long.

The annulate corolla is rose to purple with a yellow eye, and large for the size of the plant. The corolla-tube is about 0.5–1.5 cm. long and the limb 1.5–3.5 cm. diameter, with obcordate lobes notched at the apex to about half their length. Pin-eyed and thrum-eyed forms occur.

Very rare.

P. takedana Tatewaki

This rare species was first discovered in 1937 by Tatewaki. It is recorded as growing in subalpine meadows and rocky places along a tributary of the river Teshio in Hokkaido, the large island to the immediate north of the mainland of Japan.

Smith and Fletcher (7) say 'Growth begins early in the year from pinkish resting buds which are about the size of a pea.' I have obtained a plant in autumn, when the leaves were dying back. The foliage suggests that the species is a near relative to the section Cortusoides. But the bud is slender, brownish-purple and sharply pointed, about 2 cm. tall and more like a bullet than a pea, so that whether I have the true plant or not remains a matter for further observation. It should be noted that this habit of forming a brown scale-clad resting bud is restricted to species of the section Reinii. Species of the section Cortusoides do not have this development – they are completely herbaceous.

P. takedana is a more or less softly hairy perennial. The leaves, including the petiole, are 8–16 cm. long and 3–6 cm. broad. The petiole is about twice the length of the lamina. The lamina is orbicular to reniform, deeply cordate at the base and 5–7 lobed, the triangular lobes being widely toothed.

The slender scape is 8–15 cm. long and carries an umbel, or rarely two superposed umbels, of 2–3 flowers. The pedicels are 8–12 cm. long.

The corolla is white and more or less campanulate to funnel-shaped. The obovate-oblong lobes are notched at the apex and semi-erect, as with some species of Campanulaceae. The limb is 1–1.5 cm. diameter.

P. tosaensis Yatabe

P. tosaensis is a near-relative of the type species *P. reinii*, from which species it is distinguished by its longer corolla-tube, the more

shallowly notched corolla-limb and by the shape of the calyx teeth.
The leaves are 4–18 cm. long and 2–8 cm. broad. The lamina is
orbicular or broadly ovate.

The scape is 6–14 cm. tall and carries an umbel of 2–6 flowers. The
pedicels are 3–10 cm. long.

The annulate corolla is light purple. The corolla-tube is 1.5–2 cm.
long and the limb 1.5–2.5 cm. diameter with obcordate lobes notched
at the apex to not exceeding one quarter length.

The species has been in cultivation in Britain and flowered well,
but it is doubtful if it now remains.

Distribution is confined to the Japanese islands of Honshu (the
mainland of Japan) and Shikoku and Kyushu to the south-west of
Honshu.

3 THE SECTION PINNATAE

THE SPECIES OF THE SECTION

P. cicutariifolia	*P. filchnerae*	*P. ranunculoides*
P. erodioides	*P. merrilliana*	

This is a small group of Chinese species, one characteristic of which
is a much-divided efarinose leaf. There are other species which have
this same feature of the much-divided leaf, but which do not belong
to this section – *Pp. pinnatifida, incisa* and *laciniata*, for example,
which belong to the sections Muscarioides, Souliei and Farinosae
respectively.

Each of the species of this section has been found on one occasion
only, and none has ever been in cultivation. The species listed were
found in the Chinese provinces of Chekiang, Anhwei, Shensi and
East Kiangsi. While the northern province of Shensi is mountainous, Anhwei is largely a flood-plain province of central China;
and the hills of Chekiang, a coastal province of the east, are also of
lesser altitude. These species, therefore, are hardly alpine primulas.

P. cicutariifolia is a tiny annual only 2–3 cm. high. The remaining
four species are herbaceous perennials, but all more or less dwarf.
The most vigorous is the type species, *P. filchnerae*, which at flowering time has a scape about 12 cm. tall.

[48]

Plate I *P. sikkimensis* in the Wild. (Reproduced from A.G.S. *Bulletin*, Vol. 8:
'*Some Himalayan Primulas*' by G. Sherriff and G. Taylor) (p. 134)

Plate II *P. obliqua* in the Wild. (Reproduced from A.G.S. *Bulletin*, Vol. 8: '*Some Himalayan Primulas*' by G. Sherriff and G. Taylor) (p. 122)

Plate III *P. reinii* (p. 46) *Photo: Royal Botanic Garden, Edinburgh*

Plate IV *P. aureata* (p. 61) *Photo: R. C. Elliott*

Plate V *P. boothii* (p. 62)   *Photo: Royal Botanic Garden, Edinburgh*

Plate VI *P. bracteosa* (p. 63)   *Photo: Royal Botanic Garden, Edinburgh*

Plate VII *P. calderiana* (p. 67)   *Photo: R. C. Elliott*

Plate VIII *P. edgeworthii* (p. 64)   *Photo: Downward*

Plate IX *P. gracilipes* (p. 59)   *Photo: Downward*

Plate X *P. griffithii* (p. 68)

Photo: *Malby*

Plate XI *P. scapigera* (p. 64)   Photo: *Royal Botanic Garden, Edinburgh*

4 THE SECTION PYCNOLOBA

This is a monotypic section the sole member of which, *P. pycnoloba*, is described by Smith and Fletcher (6) as 'peculiar, even bizarre . . . in its inflorescence quite unlike that of any other primula'.

P. pycnoloba is a plant of hill meadows, having been found at about 2000 m. near Tatsienlu in western Szechuan. But it is apparently a rare plant, for it has seldom been found.

The species is a more or less robust, hairy perennial which spreads by means of rootbuds. It does not set seed in cultivation, but has been propagated from these rootbuds.

The distinguishing features of *P. pycnoloba* are the contracted racemose inflorescence, and the tiny corolla within the throat of a very large calyx. The corolla-tube is yellowish-green and the limb is dark red to bright violet with a yellowish-green eye.

It is doubtful if this species remains in cultivation.

5 THE SECTION OBCONICA

The Species of the Section

P. ambita	*P. filipes*	*P. parva*
P. asarifolia	*P. kwantungensis*	*P. petitmenginii*
P. barbicalyx	*P. kweichouensis*	*P. sinolisteri*
P. densa	*P. listeri*	*P. tsiangii*
P. dictyophylla	*P. obconica*	*P. vilmoriniana*
P. dumicola	*P. oreodoxa*	

This section comprises a complex assemblage of closely related plants. Some of these are not fully known and few have been in cultivation. The status of some species is, therefore, somewhat uncertain – particularly that of *Pp. petitmenginii, parva* and *barbicalyx*. These species are distinct from one another, but may prove to be forms of *P. obconica*.

Species of this section usually have a well developed rhizome, and both leaves and inflorescence are quite efarinose. The leaves have a distinct petiole, which is usually longer than the blade. The lamina is usually broadly elliptic to orbicular, cordate at the base and wavy to shallowly lobed at the margin. The scape is usually longer than the leaves. The pedicels are well developed. Both pin-eyed and thrum-eyed forms of the flower usually occur.

In China the section is most fully developed in the provinces of Szechuan and Yunnan, and thence, to a lesser degree, eastwards to Hupeh and Kweichow and southwards to Kwangtung. In the Himalaya, where again the section is not well developed, distribution extends from Nepal to Bhutan and south-eastern Tibet, and the adjoining parts of Assam, Manipur and Burma.

The usual elevation of the section is 2000–3000 m., which is too low for species of the section to prove generally hardy in Britain. The best known member of the section is the type species *P. obconica*, which the horticulturalists have developed as a house-plant.

6 THE SECTION SINENSES

The species of the section are

P. rupestris   *P. sinensis*

This is a section of two species only. *Pp. rupestris* and *sinensis* are separated from the species of adjacent sections by the characteristics of the calyx. In Sinenses this is much inflated and broad at the base. Although the adjacent species are otherwise morphologically similar, they do not hybridize, in cultivation, with the two species of this section.

The distribution of *P. rupestris* is confined to the upper valleys of the Yangtze river. It has been grown and flowered at various times since 1891, but always as a result of the importation of fresh seed, for the plants never persisted long. It does not appear to be grown at the present time.

P. sinensis, on the other hand, was never known in the wild. It is said to have first been brought to Britain 'from gardens at Canton' in about 1821. This is not a hardy species, but as a greenhouse primula it will flower in mid-winter, and there are many horticultural forms.

7 THE SECTION MALVACEA

THE SPECIES OF THE SECTION

P. bathangensis	*P. celsiaeformis*	*P. saturata*
P. blattariformis	*P. malvacea*	

[50]

The species of this section are of dwarf to medium stature, often with a basal tuft of a few farinose leaves. These are usually long-stalked, with blade orbicular and deeply cordate at the base.

The flowers are carried in a rather loose, racemose inflorescence. The colour of the flowers is rose-purple to lilac or white (*Pp. blattariformis, celsiaeformis,* and *malvacea*) or yellow (*Pp. bathangensis* and *saturata*). The maturing calyx becomes much enlarged and takes the form of a plate or wheel-like structure.

Pp. celsiaeformis and *saturata* have never been in cultivation, and are but little known in the wild. The other three species have, from time to time, been grown from seed. But the cultivated plants have never set seed and never persisted for long, although *P. bathangensis* was also propagated from root cuttings.

The species of this section appear to like drier conditions than most of the Asiatic primulas. Forrest records *P. blattariformis* as a lover of dry shade, nestling in arid rock gullies and protected by sheltering scrub. It grows at an altitude of 2000–3700 m.

Species of this section are not generally hardy.

8 THE SECTION CAROLINELLA

THE SPECIES OF THE SECTION

P. cardioeides	P. henryi	P. obovata
P. chapaensis	P. huana	P. partschiana

The species of this section are plants of the mountain forests of south-eastern Yunnan. The long-petioled efarinose leaves are few to very few and the inflorescence is a much contracted raceme.

Each of these species has been found on one occasion only, and the section is, therefore, somewhat imperfectly known.

Viable seed was never obtained and none of the species have ever been in cultivation.

The type species is *P. henryi*, and the section was named for the wife of Hemsley, who first collected *Pp. henryi, partschiana* and *obovata* in about 1896–7.

9 THE SECTION VERNALES

This European Section is omitted.

10 THE SECTION AMETHYSTINA

The Species of the Section

P. amethystina	**P. kingii*	*P. valentiniana*
*P. dickieana	*P. odontica*	*P. virginis*
P. faberi	*P. silaensis*	

KEY

Flowers various colours, not red. Corolla funnel-shaped	*dickieana*
Flowers deep red. Corolla bell-shaped	*kingii*

Species of this section have a short, slender to stout rootstock. The leaves are of a firm to fleshy texture, usually with a horny margin which is more or less remotely toothed, and they are glabrous and efarinose. With the exception of *P. dickieana* the corolla is always bell-shaped.

The species of the section are clearly related with possibly a distant relationship with the section Farinosae. In passing through the sections of the genus we are therefore now moving away from the aggregates of Cortusoides and Obconica.

The distribution of the section is entirely Asiatic, and confined to the eastern Himalaya and the mountains of south-eastern Tibet, south-western China and northern Burma.

The species of this section are particularly intractable in cultivation. *Pp. amethystina, dickieana, kingii* and *odontica* have been flowered, and *P. valentiniana* has been raised but not flowered. Of these species, only *P. kingii* remains in a very few private gardens.

P. dickieana Watt

This species is a perennial with a short rootstock which is clothed above with many small scale-leaves. The plant has a tufted habit, with leaves, including the petiole, 2–7 cm. long and 0.3–1 cm. broad. The winged petiole is at first short but may, later, equal the blade. The lamina is elliptic-obovate to oblanceolate, more or less acute at the apex, cuneate at the base, and efarinose. The margin is more or less entire, or minutely and remotely sharp-toothed.

The scape is 8–20 cm. tall and carries a single umbel of 1–6 drooping flowers. The stout pedicels are 2–7 mm. long, but up to 2.5 cm. in fruit and then becoming erect.

The corolla may be yellow, white, mauve, violet or purple, usually with a yellow eye. There is no annulus. The length of the flower is about 2 cm., with the corolla-tube nearly 2 times the calyx and hairy at the mouth and on the inner surface. The limb is 2–3 cm. diameter, with lobes obcordate to elliptic or oblong and deeply divided at the apex.

P. dickieana grows in moist, alpine meadows at an elevation of 4000–5000 m. Its distribution includes Sikkim, Bhutan and south-eastern Tibet and it also occurs just within northern Burma and north-western Yunnan.

This distinctive species has been re-introduced into cultivation by seed sent back from Nepal by Beer in 1975, and has germinated well in a number of gardens.

P. kingii Watt (Illustration p. 19)

This species is a perennial plant with a short, fairly stout rootstock. The efarinose leaves, including the short, broadly-winged petiole, are 2–6 cm. long and 0.5–1 cm. broad, elliptic-lanceolate, acute at the apex and tapering below. The more or less horny margin is entire, faintly lobed or remotely toothed.

The scape is 1–20 cm. high, but up to 30 cm. in fruit, and carries a single umbel of 2–10 semi-pendent flowers. The pedicels are 5–15 mm. long, scarcely lengthening in fruit.

The dark claret, bell-shaped corolla is 1.5–1.8 cm. long and 1.5 cm. diameter at the mouth. The petal lobes are elliptic or oblong, retuse at the apex and entire at the margin. The annulus is absent, or only faintly developed.

This is a plant of wet meadows from upwards of 4000 m.

11 THE SECTION CUNEIFOLIA

The Asiatic species of the section are:

P. cuneifolia   *P. nipponica*

The position of this section was long a matter of doubt. In 1928 Smith and Forrest (4) noting certain fariniferous glands at the apex of the scape and on the pedicels felt that the section was 'somewhat isolated, with its nearest kindred to be found among the Farinosae'.

A few years later, it was noted by R. E. Cooper of the Royal

Botanic Garden, Edinburgh, that the young leaves of cultivated plants were involute. This proved to be the missing link – Cuneifolia should be placed close to the European section Auricula. It was subsequently appreciated that morphological and cytological data fitted this interpretation. Cuneifolia as entered on the sectional list (p. 2) should therefore be much closer to Auricula.

The distribution of *P. cuneifolia* ranges from north-eastern Siberia to Japan, the Aleutian Islands and the coast of Alaska. *P. nipponica* is confined to Japan.

P. cuneifolia Ledeb. (Illustration p. 20)

This is a more or less glabrous perennial with a short stout rhizome. The leaves including the petiole are 1.5–8 cm. long. The lamina is 1–3 cm. long and 0.5–2.5 cm. broad, oblanceolate to obovate-cuneate and tapering into the membranously winged petiole. The margin is coarsely toothed, often in the upper half only.

The scape is 6–30 cm. tall and carries an umbel of 1–9 flowers. The pedicels are 4–20 cm. long.

The annulate corolla is rose-red or crimson with a yellow eye. The corolla-tube is about 0.5 cm. long and the limb is 1–2 cm. diameter, with obcordate lobes shallowly notched at the apex.

P. cuneifolia has flowered well in cultivation, but does not long persist in the absence of seed.

12 THE SECTION PETIOLARES

THE SUBSECTIONS, GROUPS AND SPECIES OF THE SECTION

SUBSECTION PETIOLARIS-SONCHIFOLIA

GROUP VERA

*P. cunninghamii	*P. gracilipes	*P. sessilis
*P. deuteronana	P. hoffmanniana	P. sulphurea
P. drummondiana	*P. petiolaris	

GROUP SCAPIGERA

*P. aureata	*P. bracteosa	P. moupinensis
*P. boothii	*P. irregularis	*P. scapigera

GROUP TALIENSIS

P. comata	P. taliensis

GROUP ODONTOCALYX

P. euosma	P. odontocalyx	P. sinuata

GROUP HOOKERI
P. hookeri   P. vernicosa

GROUP EDGEWORTHII
*P. edgeworthii

GROUP SONCHIFOLIA
P. chamaedoron   P. chionogenes   *P. sonchifolia
P. chamaethauma   P. chionota   *P. whitei

SUBSECTION CHARTACEA
P. chartacea   P. petelotii   P. veitchiana
P. lacerata

SUBSECTION TONGOLENSIS
P. pulchra   P. tongolensis

SUBSECTION GRIFFITHII
*P. calderiana   P. jucunda   *P. strumosa
P. calthifolia   P. laeta   P. tanneri
*P. griffithii   *P. nepalensis   *P. tsariensis
P. hilaris

SUBSECTION DAVIDII
P. breviscapa   P. epilosa   P. klaveriana
P. coerulea   P. esquirolii   P. leptophylla
P. crassa   P. fagosa   P. ovalifolia
P. davidii   P. hylobia

KEY

1 Scape absent or less than 4 cm. at flowering time.
     Winter bud or rosette   2
  Scape 5 cm. or more at flowering time.
     Herbaceous (Subsect. Griffithii)   15

2 Scape absent (Group Vera)   3
  Scape less than 4 cm. at flowering time,
     elongating later (Group Scapigera)   7

3 Flowers annulate   4
  Flowers exannulate (or faintly annulate)   5

4 Petals 3-toothed. Corolla with white zone; tube
     hairless within   gracilipes
  Petals irregularly toothed. Corolla with no white
     zone; tube hairy within   deuteronana

5 Petals deeply notched   cunninghamii
  Petals rounded or toothed   6

[55]

6 Petals 3-toothed	*petiolaris*
Petals rounded with 1 tooth	*sessilis*
7 Flowers yellow, pink or purple. Winter foliage a tight rosette	8
Flowers blue. Winter resting bud (Group Sonchifolia)	14
8 Flowers annulate	9
Flowers exannulate	11
9 Flowers yellow	*aureata*
Flowers pink, lilac or purple	10
10 Petals toothed. Leaves at tip of scape. Farina white	*bracteosa*
Petals wavy or lobed. No scape-leaves. Farina yellow	*irregularis*
11 Leaves of two types, according to season	*edgeworthii*
Leaves of one type only	12
12 Leaves at tip of scape	*bracteosa*
No scape-leaves	13
13 Calyx lobes triangular	*boothii*
Calyx lobes long-pointed	*scapigera*
14 Flower with white 'eye'	*whitei*
Flower with yellow 'eye'	*sonchifolia*
15 Flowers yellow	16
Flowers purple or maroon	17
16 Calyx lobes lanceolate	*nepalensis*
Calyx lobes ovate	*strumosa*
17 Petals deeply notched	*tsariensis*
Petals shallowly notched	18
18 Leaves oblanceolate; tapering at base	*calderiana*
Leaves ovate; cordate or sagittate at base	*griffithii*

This is the second largest section of the genus, with approximately sixty species. Of these, 18 species and some varieties and hybrids are in cultivation.

The composition of this section was long a matter of argument. Various morphologically similar species were grouped round the type species, *P. petiolaris*, and subsections and groups were created. It was subsequently realized that these kindred species had one feature in common: a capsule the walls of which, at maturity, completely crumble away to scatter the seed. There are no horny apical teeth and persistent capsules among the Petiolarids. This feature of the crumbling capsule is characteristic of all species of the section. There is, however, the possibility that this characteristic is too wide, and by reason of its use some species may have been placed within the section which should be elsewhere. Only further examination of plants, both in the wild and in cultivation, will finally establish the true status and position of some of these species, many of which are but imperfectly known and difficult to cultivate.

The distribution of the section extends from the Himalaya to the mountains of western China, south and south-eastern Tibet and north-eastern Burma. The area involved is continuous, but the various subsections and groups are more or less localised. Thus the subsections Davidii and Chartacea, and the groups Taliensis and Odontocalyx, are almost entirely restricted to western China. The subsection Griffithii is from the eastern Himalaya and south-eastern Tibet. The group Sonchifolia is from the eastern Himalaya, Tibet and western China and *P. edgeworthii* is confined to the western Himalaya. The groups Vera and Scapigera are the most widespread.

While there is no definite gap between species of the section from the Himalaya and those of western China, the degree of kinship is somewhat limited, for there are no wide-ranging species common to both areas as is the case, for example, with certain species of the section Nivales.

The plants of this section are perennial. For the genus, they are of average stature or a little less. Petiolaris-Sonchifolia, the largest sub-section, may be taken as the median type. The root system is fibrous to more or less fleshy, the latter usually associated with the more vigorous types, and there may be a long, thin much-branched rhizome. Some species, e.g. *P. griffithii*, are completely herbaceous, while other species, e.g. *P. sonchifolia*, persist as a stout fleshy resting bud which is usually more or less farinose and enclosed by scale-leaves. With yet other species, perhaps the majority, the summer foliage, typical long-petioled, spoon-shaped leaves with a toothed margin, dies away to expose a crown of much-contracted, strongly silver or yellow-farinose leaves. These slowly expand, very early (or late) in the year to expose the tightly packed, sessile flower buds.

The development of the scape is also very variable within the section, but constant within the species. Three types of development may be identified. In the first type, group Vera and *P. edgeworthii*, the flowers are held singly on a thin flexuous pedicel at the level of a much-contracted winter foliage, and the pedicels do not further elongate in fruit. In the second, groups Scapigera and Sonchifolia, the flowers are carried in an umbel at the head of a scape which is slow to develop and much contracted at the time of flowering, but subsequently elongates and develops rapidly in fruit. In the third type, the more vigorous Petiolarids, subsections Griffithii and Davidii, the scape develops early. The flowers are carried in an umbel well above the leaves and the scape develops still further when fruiting. There is no sharp demarcation between scape development of the second and third types.

The corolla may be annulate or exannulate, and pin-eyed and thrum-eyed flowers occur. Colour is rose to deep purple, or occasionally yellow or white. There is a considerable amount of variation in the form of the corolla-lobe of species of the various subsections and groups, but persistence of this character within individual species. The form of the corolla-lobe can therefore be used as a diagnostic feature.

SUBSECTION PETIOLARIS-SONCHIFOLIA
GROUP VERA

P. cunninghamii King

The distribution of this species is confined to the eastern parts of Sikkim and the contiguous Chumbi valley in Tibet, where it grows at 3300–4300 m. on open hillsides or among rhododendron, often on gravel or turf. It is closely related to two other Sikkim species, *Pp. deuteronana* and *gracilipes*, but in these two species the corolla is quite different from that of *P. cunninghamii*, in both size and form of the lobe. It is reported as flowering between September and February. Smith and Fletcher said, in 1944 (5), that the species had never been in cultivation. But a plant is now occasionally offered under this name, and the flower-buds are colouring now in mid-November, on plants grown in the open.

P. cunninghamii is a dwarf plant with whitish-yellow farina. The leaves form a compact rosette and are 1.5–11 cm. long and 1–4 cm. broad, oblong-spathulate to oblong-obovate, rounded at the apex and tapering into the winged petiole, which is almost absent on the inner leaves. The margin is irregularly but finely toothed, and the lamina more or less glabrous and efarinose at maturity.

The scape is absent. There are many pedicels. These are about 1 cm. long, thin and wiry, and each carries a single flower.

The corolla is purplish-pink to a beautiful pale lavender blue with a mustard-yellow throat which forms a five pointed star, surrounded by a white pentagonal area. The corolla-tube is about 1 cm. long and the limb 1–1.5 cm. diameter. The lobes are obcordate and deeply notched at the apex.

P. deuteronana Craib

This species was first found in 1848 by Hooker on the Singalila range, which reaches down from Kanchenjunga and there forms the marches between Sikkim and Nepal.

This is a dwarf, more or less efarinose plant with a short, stout rhizome from which descend the thick fleshy roots.

The leaves are 1.5–5 cm. long and 1–2 cm. broad, oblong-obovate to oblong-ovate, tapering at the base into the much reduced petiole and irregularly but finely toothed at the margin.

The scape is absent, and the pedicels 1–2 cm. long, each carrying a single flower.

The corolla is pale purple with a yellow annulus. The corolla-tube is about 2 cm. long and the limb 1.2–3 cm. diameter with oblong-obovate lobes irregularly and shallowly lobed or toothed at the margin, or thrice-lobed at the apex. This is the only species of the section which has shaggy hairs within the corolla-tube.

The species was reported by Smith and Fletcher (5) as not being in cultivation, but has now probably been established in one or two botanic and private gardens.

P. gracilipes Craib (Plate IX)

This species is found from Nepal and Sikkim to Bhutan and southern Tibet. Watt describes it on the Singalila Range, which forms the northern part of the frontier between Nepal and Sikkim, as very frequent and in full flower in May, although the flower buds, often with much farina, may be showing in November.

This is a rather dwarf plant with a branched rhizome, the stoutness of which depends on the vigour of the particular plant. The leaves, including the petiole, are 4–15 cm. long and 1–5 cm. broad. They are efarinose, grow in a compact cluster and are very much contracted in winter. The lamina is oblong-spathulate or oblong-obovate to elliptic, rounded at the apex and tapering into the broadly-winged petiole, or more or less rounded at the base with a distinct petiole. In some plants, the midrib of the petiole is red. The margin of the leaf is irregularly toothed.

The scape is absent. The long thin pedicels are 1–6 cm. long and each carries a single flower.

The annulate corolla is bright pink-purple with an orange-yellow eye, which is surrounded by a narrow zone of white. The corolla-tube is 2–3 cm. long and the limb 1.5–3 cm. diameter, with broadly obovate lobes usually divided into three toothed segments.

The closely related species are *Pp. cunninghamii* and *deuteronana*. The former can be distinguished by the shallowly notched corolla lobes and the latter by the shaggy hairs within the corolla-tube.

P. petiolaris Wall.

This is a small efarinose plant with a thin rhizome from which descend many fibrous roots.

The leaves grow in a loose rosette, which is much contracted in winter. The inner leaves, including the reduced petiole, are 3–5 cm. long and 1–1.5 cm. broad, spathulate to oblong-spathulate, rounded at the apex, tapering at the base into the winged petiole, and toothed at the margin. The outer leaves are 2–3 times as long as the inner leaves.

The scape is absent. The many pedicels are 2–5 cm. long and each carries a single flower.

The corolla was originally described by Smith and Fletcher as 'pale blue'; but a plant later introduced by Ludlow, Sherriff and Hicks was 'magenta-purple' (*Bulletin* Vol. 30 p. 231). The corolla-tube is 1–2 cm. long and the limb 1.5 cm. diameter with obovate lobes which are usually divided into three toothed segments.

The true species is confined to Nepal.

P. sessilis Royle ex Craib

This is a species of the north-western Himalaya. Other species of the area are *Pp. edgeworthii*, *drummondiana* and *sulphurea*. *P. sessilis* is readily distinguished from these species, but there may be confusion with the central Himalayan *P. petiolaris* from Nepal.

This species has about the same growth as *P. petiolaris*. The scape is absent and the flowers are held singly on thin pedicels 2–10 cm. long. The corolla is pale mauve with a yellow eye, which is surrounded by a narrow band of white. The lobes of the limb are obovate to oblong-obovate, and rounded and suddenly acute or with a small broad tooth at the apex.

A rare plant in cultivation.

GROUP SCAPIGERA

P. aureata Fletcher (Plate IV)

This species has silvery-white farina which is absent during the growing period but is strikingly developed in winter, when the leaves form a much-contracted, tight rosette.

The leaves, including the petiole, are 4–10 cm. long and 1–3 cm. broad, rounded at the apex and tapering at the base into the broadly-winged petiole. The margin is deeply and irregularly toothed, and during the growing season the midrib of the petiole is coloured beetroot-red.

The scape is 5–10 mm. long, obscured by the leaves, and carries an umbel of 3–10 flowers. The pedicels are up to 2 cm. long.

The corolla-tube is 1.5–3 cm. long and the limb 2.5–4 cm. diameter, with broadly obovate and rather fleshy lobes which have three deeply-cut teeth at the apex, or are irregularly toothed. In colour the petals are a pale creamy-yellow with a flush of deep chrome-yellow to orange extending from the mouth of the tube to halfway along the lobes. *P. aureata* is a magnificent plant, but still rather rare.

This species was first known only in cultivation. In 1935, seed of *Swertia purpurea* was sent by the Lloyd Botanic Garden, Darjeeling, to Edinburgh, and in the pan of seedlings subsequently raised was a rogue Petiolarid primula, which flowered as *P. aureata*. It was first seen in the wild by Mrs. D. Proud, who found it growing at about 3800 m. under overhanging rocks on the Gandaki-Kosi watershed in central Nepal, in May 1952.

The second introduction of *P. aureata* was also quite unintentional. During the 1954 expedition to Nepal, a flowerless Petiolarid primula was collected under Stainton, Sykes and Williams No. 9322. This was sent back by air and when flowered at Edinburgh, again proved to be *P. aureata*. It was found growing in rock crevices at about 4000 m., on the Lamjung Himal.

L. H. J. Williams (Stainton, Sykes & Williams) says, of the Petiolarids, that they are confined to the alpine and subalpine regions. Many of them begin to flower in November and, with intermittent snow-cover, sometimes continue to flower until May. This is the reason why some of them have proved difficult to find and to understand, for there are few collectors in the field so early in the year. In their native habitat they grow in shade, often on vertical banks and in rock crevices, where there is abundant moisture during the growing season.

P. aureata was again found in 1974 by Syms, in the Langtang valley, to the north of Khatmandu.

P. aureata forma Fletcher

It is thought that this species was introduced from Nepal by Colonel Bailey in 1935. It was subsequently submitted to Dr. H. R. Fletcher, who thought the plant to be a form of *P. aureata*, an opinion with which he was apparently not entirely satisfied, but the plant is still grown and exhibited as *P. aureata forma*.

This form is similar in growth to the species, but perhaps of less vigour, and the leaves more finely cut.

The flowers are smaller than those of *P. aureata* proper, of a more or less uniform pale creamy yellow, with the central zone of colour much reduced both in area and in intensity and the lobes more deeply toothed at the apex.

There is a certain amount of variation in the colour and form of the flower. But whether this is due to propagation by seed-raising, or to cultural treatment, is not clear.

P. boothii Craib (Plate V)

P. boothii is rather dwarf. It is efarinose. The leaves, including the petiole, are 3–15 cm. long, broadly spathulate to elliptic-spathulate, rounded at the apex and irregularly toothed on the margin. The base of the inner leaves tapers into the short, broadly winged petiole, but the base of the outer leaves is more or less truncate, and there is a distinct petiole which may equal the lamina. The midrib of the petiole is tinged red.

In winter, the foliage contracts into a somewhat loose, more or less upright, bunch of leaves, the underside of which is a uniform pinkish-brown.

The scape at flowering time is zero to 2 cm. long, but develops later, and carries an umbel of up to 25 flowers. The pedicels are 1.5–5 cm. long.

The corolla is pale pinkish-purple with a yellow eye which is surrounded by a narrow white band. The corolla-tube is 1–2 cm. long and the limb 2–2.5 cm. diameter, with broadly obovate lobes which may be wavy at the margin or more deeply toothed.

[62]

P. boothii is a rare species, and not often found. It appears to be confined to eastern Bhutan.

P. bracteosa Craib (Illustration p. i. Plate VI)

This species is sparsely white-farinose. The leaves, including the petiole, are 5–16 cm. long and 2–6 cm. broad, spathulate to obovate-spathulate, rounded at the apex and tapering at the base into the broadly-winged petiole. The leaves are sharply and irregularly toothed at the margin.

The scape at first is very short and obscured by the foliage, but later thickens and elongates up to 25 cm., and carries an umbel of few to many flowers. The pedicels are 1.5–4.5 cm. long.

The corolla is pale pinkish-lilac with a large yellow eye, which is surrounded by a thin white ring. The corolla-tube is 1.5–3 cm. long and the limb 2.5–3 cm. diameter. The obcordate lobes are quite separate and well reflexed, and have a number of shallow teeth at the apex.

At late-flowering and fruiting time, vegetative buds develop at the top of the scape and these grow into stalked leaves up to 9 cm. long. If the vegetative buds are pegged down after the manner of a straw-berry-runner, then they will root in about three weeks and can be lifted and potted singly after a further two weeks. The development of a vegetative bud is not a constant feature of the species.

This is a plant of the Assam Himalaya, Bhutan, and possibly as far west as Sikkim.

P. irregularis Craib

This species has leaves in a dense rosette, at first covered below with copious yellow farina, but this quickly disappears.

At flowering time the outer leaves, including the petiole, are 8–12 cm. long and about 3 cm. broad (about three times the dimen-sions of the inner leaves) and at fruiting time up to 18 cm. long and 6 cm. broad. The lamina is oblong-elliptic to elliptic-obovate, rounded at the apex and cuneate or attenuate at the base. The margin is deeply toothed or sometimes shallowly lobed and toothed. The petiole is rather shorter than the lamina.

The scape at flowering time is 1–2 cm. long, farinose, and quite obscured by the compact rosette of leaves. The inflorescence is an umbel of 12–20 flowers. In fruit, the scape elongates to about 13 cm. The pedicels are 1.5–3 cm. long.

The annulate corolla is a bluish-purple. The corolla-tube is 1.5–2.5

[63]

cm. long and the limb 1.5–2.5 cm. diameter, with broadly oblong-ovate lobes which are wavy or lobed at the apex.

This species was first found on the Singalila Range in the west of Sikkim. It also extends across the eastern frontier and into the Tibetan valley of Chumbi. It is found at 3000–4600 m. and is fairly frequent. It is said to flower in April and May, but in cultivation it can be more or less perpetual flowering.

P. irregularis is rare in cultivation.

P. scapigera Craib (Plate XI)

P. scapigera is efarinose. The leaves, including the petiole, are 3–15 cm. and are fully developed at flowering time. The lamina is 2–7 cm. long and 1–4 cm. broad, oblong-spathulate to ovate, rounded at the apex, attenuate to cordate at the base and coarsely and irregularly toothed at the margin.

The scape at flowering time is usually less than 4 cm., and obscured by the leaves. The inflorescence is a many-flowered umbel at the head of the scape, which lengthens during flowering and at fruiting equals the leaves.

The corolla is pink with a yellow eye surrounded by a ring of white. The corolla-tube is 1.5–2 cm. long and the limb 2.5–3 cm. diameter with broadly obovate lobes which have two or three teeth at the apex, or are irregularly toothed.

P. scapigera occupies a limited area at 2300–3700 m. on the western frontier of Sikkim and from there probably extends into eastern Nepal.

GROUP EDGEWORTHII

P. edgeworthii Pax (Illustration p. 18. Plate VIII)

This species displays, to a marked degree, the two-phase pattern of growth adopted by many of the Petiolarid primulas. In autumn, the long-spathulate green leaves of summer are replaced by much contracted foliage covered with strikingly beautiful silvery-white farina. The winter foliage forms as a tight rosette or crown, at the heart of which lie the flower buds of spring.

The leaves, including the petiole, are 3–17 cm. long (the smaller dimension in the better forms) and 1–6 cm. broad. At flowering-time the profile is spathulate to obovate-spathulate, rounded at the apex, tapering at the base into the short, winged petiole, and toothed at the margin. At late flowering and fruiting time the foliage is suc-

[64]

Plate XII *P. sonchifolia*—Resting Bud in Winter (p. 65)   *Photo: F. H. Fisher*

PRIMULA SONCHIFOLIA

Plate XIII *P. sonchifolia* (p. 65)

Photo: *R. C. Elliott*

Plate XIV *P. forrestii* (p. 74)   Photo: *H. S. Wacher*

Plate XV *P. algida* (p. 84)   *Photo: O. Schwarz*

Plate XVI *P. clarkei* (p. 96)   *Photo: R. C. Elliott*

Plate XVII *P. inayatii* (p. 92)

Plate XVIII *P. warshenewskiana* (p. 98)

Plate **XIX** *P. denticulata* (p. 99)

Photo: R. C. Elliott

Plate XX *P. vialii* (p. 105)

ceeded by scarcely farinose leaves which are ovate with a rounded apex, and base more or less truncate, with a petiole 1–3 times the length of the lamina.

The scape is absent, or up to 1.5 cm. long and obscured by the foliage, and carries an umbel of few to many flowers. The pedicels are 2–8 cm. long.

The colour of the corolla is more or less pale mauve. In some forms the colour is rather stridently dark and inky, but there is a very beautiful clear medium blue form, with dark mustard-yellow eye surrounded by a white band. There is also a very beautiful white-flowered form under the style of *P.e. alba.*

The corolla-tube is 1.5–2.5 cm. long and the limb 2–3 cm. diameter, with broadly obovate, irregularly and sharply toothed lobes.

P. edgeworthii is a species of the western Himalaya, in particular Kumaon, where it grows at upwards of 3000 m. The dimorphic tendency of the leaves, as displayed by the totally dissimilar form of the leaves of summer and winter, long gave rise to confusion as to the identity and naming of the species, which at one period was also known as *P. winteri.*

GROUP SONCHIFOLIA

P. sonchifolia Franch. (Plates XII and XIII)

The winter-resting form of this species is a large fat bud which is protected by more or less horny overlapping scale-leaves coloured greenish or yellowish-brown.

When the bud-scales open in early March, the flowers nestle at the heart of a rather open rosette of ovate leaves, which are then raggedly toothed at the margin and speckled with random blotches of whitish farina. The leaves are later 20 cm. long and 5 cm. broad and, in fruit, up to 35 cm. long and 12 cm. broad, oblong to obovate-oblong, rounded at the apex and tapering at the base into the broadly-winged petiole and coarsely and raggedly toothed at the margin. At this time, if provided with adequate shade and moisture, *P. sonchifolia* is a beautiful and vigorous foliage plant.

The scape, at first, is obscured by the foliage, but later develops and reaches 30 cm. in fruit, and carries a single umbel of 3–20 flowers. The pedicels are 1–2 cm. long.

The corolla is lavender-blue or purplish blue to bright blue or intense indigo-violet, or sometimes white, with a dark mustard-yellow 5-lobed eye surrounded by a narrow pentagonal whitish

[65]

band. The corolla-tube is 1–3 cm. long, about 2–4 times the length of the calyx, and the limb is 1.5–2.5 cm. diameter, with broadly obovate to suborbicular lobes and few to many fringing teeth at the apex or occasionally entire or slightly notched.

The capsule of all the Petiolarid primulas readily crumbles away and seed is washed out in a newly matured or 'green' condition. With *P. sonchifolia*, the viability of seed decreases rapidly with departure from this green condition and all the early attempts to introduce this species by seed conveyed by land and sea were doomed to failure. Live, dormant plants were eventually transported in a hollow bamboo container, which was kept under refrigeration during the sea voyage; and at the present time fresh seed can be successfully transported by air.

P. sonchifolia is a plant of open mountain meadows on the verge of the snow-line. It grows at 3300–4600 m. on the western ranges of Yunnan and Szechuan and on the frontiers of south-eastern Tibet and Burma. It is a plant of similar growth and habit to *P. whitei*, but it is a more vigorous plant and does not require the overhead winter-protection demanded by most of the Petiolarid primulas. When the winter contraction takes place the root-system also seems to diminish to a few fibrous hairs. The only difficulty I find with the plant is to stop the birds scratching out the buds when, in this phase, they have insufficient anchorage. The answer seems to be to net the plant or, preferably, to protect the buds with small, close-spaced pieces of rock.

P. whitei Fletcher

With the exception of *P. boothii*, the winter-resting form of all the Petiolarids so far described has been a compact rosette of much-contracted leaves, often with farina much more widely developed than in the summer growing season. With species of the group Sonchifolia, the winter resting condition takes the form of a large fat bud, which is protected by scale-leaves. In *P. whitei* these are yellow-farinose and overlapping.

At flowering time the leaves, including the petiole, are 2–10 cm. long and 1–3 cm. broad, spathulate to oblong-spathulate, obtuse or rounded at the apex, tapering at the base into the very short petiole, and irregularly toothed at the margin. The leaves at first are farinose, but later efarinose above and only sparsely farinose beneath. At fruiting time, the foliage is up to 20 cm. long and 5 cm. broad, with a well-developed, winged petiole which may almost equal the lamina. The petiole and midrib of the lamina are then stained dark red.

The scape at first is quite short and obscured by the foliage, but later, carrying few to many flowers, develops conspicuously and may then be 1–8 cm. long, but up to 18 cm. in fruit. The pedicels are 2–4 cm. long.

The slightly annulate corolla is pale or dark blue with a white eye which is occasionally tinged greenish-yellow in the centre. The corolla-tube is 1.5–2 cm. long, about twice the length of the calyx, and the limb is 2–3 cm. diameter with obovate lobes with three deeply-cut teeth at the apex.

P. whitei was long confused with *P. bhutanica*, the latter being held to be a separate species. There are indeed still some who would argue that *P. bhutanica* has specific status, but the balance of opinion appears to regard *Pp. whitei* and *bhutanica* as one and the same plant, and *P. bhutanica* is here included with *P. whitei*.

P. bhutanica was first found by Kingdon-Ward in the Assam Himalaya in 1935. Its distribution also extends into Bhutan and south-eastern Tibet, where it grows at 3000–4300 m. in coniferous and mixed forest, under rhododendron and on damp mossy banks.

The species is in cultivation, but not readily available.

SUBSECTION GRIFFITHII

The species of this subsection are completely herbaceous. There is no sign of growth above the soil surface during the period of winter dormancy.

The leaves are variable. With *P. calderiana* the leaves are strap-shaped and rather like a less-vigorous Nivalid. With other species of the subsection the green-crinkled leaves are more or less spathulate as with most of the Petiolarids.

The scape is already well-developed at flowering-time, and elongates in fruit.

P. calderiana Balf.f.et Cooper (Plate VII)

P. calderiana is a species of Sikkim and the eastern Himalaya. It is a rather variable plant, due largely to its considerable altitudinal range, which is 3700–5000 m. There is even a stemless form, which is known as var. *acaulescens*.

P. calderiana has short, fleshy roots which can be teazed apart rather after the fashion of gentian thongs. It is described, on good authority (5), as a disagreeably smelling plant, but I have not found this to be the case.

The leaves, including the petiole, are 5–30 cm. long and 1–6 cm.

broad and usually fully developed at flowering time. They are more or less efarinose, particularly on the upper surface. The lamina is spathulate to oblanceolate, rounded at the apex and tapering below into the broadly winged petiole. It is generally not possible to distinguish between lamina and petiole.

The scape is 5–30 cm. tall, elongating considerably in fruit, and carries an umbel of 2–25 flowers. The pedicels are 0.5–3.5 cm. long, and may be upright or drooping.

The annulate corolla is purple to maroon with a slightly darker area at the centre and a yellow, pentagonal eye. The corolla-tube is 1–2 cm. long, about twice the length of the calyx, and the limb 1.5–3 cm. diameter with very broadly obcordate, overlapping lobes.

P. griffithii (Watt) Pax (Plate X)

This, the type species of the subsection, has a short, stout rootstock. The leaves, including the petiole, are 6–30 cm. long and 2–6 cm. broad. The lamina is ovate to almost sagittate, with occasional patches of pale yellow farina below. The base of the lamina tapers into the broadly winged petiole, or is rounded, or even more or less cordate. The margin is variously toothed. The length of the petiole is rather less than that of the lamina.

The scape is 10–20 cm. long at flowering, but up to 45 cm. in fruit, and carries an umbel of 5–12 flowers. The pedicels are 1–2 cm. long.

The annulate corolla is deep purple with a yellow eye. The corolla-tube is 1–2 cm. long, which is 2–3 times the calyx. The limb is 2–2.5 cm. diameter, with obcordate lobes shallowly notched at the apex.

While *P. griffithii* is exceedingly rare in cultivation, one nurseryman offers a *P. griffithii* hybrid of which *P. griffithii* was the seed parent, pollinated by *P. strumosa*. The original cross was made at Keillour Castle. Since these hybrids are propagated by seed, the colour of the corolla may be anything from purple to yellow. A plant should therefore be seen in flower before making a purchase, or seed may be collected and the seedlings selected according to colour after they have flowered. I have a plant of this cross which is a good, deep violet-blue, free-flowering, and of moderate vigour, and must be fairly close to the true *P. griffithii*. This plant has been successfully divided after flowering, and seems to be a better doer than the *P. strumosa* hybrid described below.

P. griffithii is a plant of western Bhutan, the south Tibetan valley of Chumbi which drives a wedge between Bhutan and Sikkim, and probably also extends to the west into Sikkim.

P. nepalensis W.W. Sm.

This species is a plant with a short rhizome, from which grow the numerous fleshy roots. The leaves, including the petiole, are 10–25 cm. long. The lamina is 6–10 cm. long and 3–7 cm. broad, elliptic to ovate-cordate, rounded to acute at the apex, cordate to cuneate at the base, deeply and irregularly toothed at the margin and always efarinose on both surfaces.

The stout scape is 20–30 cm. tall and carries an umbel of 6–12 flowers. The pedicels are 1–2 cm. long.

The corolla is yellow, with a large annulus. The corolla-tube is 1–2 cm. long and the limb 1.5–2.5 cm. diameter, with oblong to narrowly-obovate lobes which are shallowly notched at the apex.

This species was originally collected, in 1930, by Major Lal Dhwoj at 3700–4000 m. in Nepal. It may occur in a very few private gardens, but is grown with some success at Edinburgh and may become more generally available.

P. strumosa Balf.f. et Cooper

This species is a plant with a short rootstock, from which arise the numerous, fleshy roots. The leaves, including the petiole, are 3–23 cm. long and 1.5–6 cm. broad, but up to 30 cm. in fruit, oblanceolate to obovate, obtuse to rounded at the apex, narrowing rapidly at the base and sharp or round-toothed at the margin. The lamina may be efarinose, or sometimes with copious yellowish farina below. The winged petiole is more or less equal to the lamina.

The stout scape is 7–35 cm. tall and carries a few to many flowered umbel. The pedicels are 1–2 cm. long.

The annulate corolla is yellow with an orange-yellow eye. The corolla-tube is 1–3 cm. long, which is 2–3 times the calyx, and the limb 1.5–2 cm. diameter, with orbicular to oblong lobes which are slightly notched at the apex.

This species was discovered in 1915 by Cooper, growing at 4700 m. in Bhutan in open turf amongst boulders. It has also been found in Nepal, growing in company with *P. nepalensis*, a closely related species from which it may be distinguished by the shape of the leaves, which are always efarinose in *P. nepalensis*, and by details of the calyx and corolla.

P. strumosa might be found in one or two private gardens. In 1975 seed was sent back from Nepal by Beer, but the viability of the seed of the Petiolarid primulas is notoriously doubtful.

One British nurseryman offers a *P. strumosa* hybrid which, like the *P. griffithii* hybrid, originally came from Keillour Castle. The *strumosa* hybrid had *P. strumosa* as the seed parent and was pollinated by *P. griffithii*. Both the *Pp. griffithii* and *strumosa* hybrids are plants with crinkly green leaves. The former appears to be the better doer, the latter more delectable to some leaf-eating pest.

It is always worthwhile acquiring the *P. strumosa* hybrid. The plant purchased may not be near *P. strumosa*, but plants raised from one's own seed may run near to the true species, and that is a magnificent plant.

P. tsariensis W.W.Sm.

The distribution of this species is south-eastern Tibet to central Bhutan, where it grows at 3500–5000 m. 'on moist hill slopes (and) on the banks of streams, from near the snow down to the edge of the bamboo forest' (5).

P. tsariensis was first collected by Kingdon-Ward in 1935 and subsequently by Ludlow, Sherriff and Taylor in 1936–8. As with many Petiolarid primulas, attempts to raise plants from seed were not successful. But living plants were also acquired and sent back by airmail, and these were established and flowered. The species remains in cultivation, but is rare and not generally available.

With the species occurring over the considerable altitudinal range of 1500 m., there is much variation in vigour, growth and form. The best forms are described as 'a magnificent Primula' (5).

P. tsariensis has a short rhizome. The leaves are dark green, sometimes tinged with purple, efarinose. The lamina is 2–4 cm. long and 1.5–2 cm. broad, elliptic to ovate-lanceolate, rounded to obtuse at the apex, subcordate to broadly cuneate at the base and round-toothed at the margin. The winged petiole is more or less equal to the lamina.

The scape is 2–12 cm. tall, often stained purple, and carries a single umbel of 1–8 flowers. The pedicels are 0.5–2 cm. long.

The annulate corolla is bright pinkish-purple to rich blue-purple, sometimes purplish-yellow and occasionally white, but always with a yellow eye. The corolla-tube is 1–2 cm. long which is 1–2 times the length of the calyx. The limb is 1.5–3 cm. diameter, with broadly obovate lobes deeply notched at the apex.

A Note on the Cultivation of the Petiolarid Primulas

The Petiolarid primulas grow with greater vigour when planted in the open ground – at least in the north of England. They are very difficult in the south. They like a leafy soil and must have shade, quite deep shade, for they are mostly woodland plants. A shaded area also has the advantage that it can be kept moist in summer, whereas water will run off a sun-baked slope.

When dealing with the small-scale cultivation of Nivalid primulas (Section 27) it is suggested that species of that group may be grown in the bottom of a hollow or depression formed in the lee of a beech hedge. The sides of such a hollow can be built up with small pieces of rock, say 8 in. across, and the Petiolarids interplanted. The crown of a plant should never be flat, but always have as much inclination as possible so that water will run out of the crown. The roots of each plant should be under a stone and the top surface of the stone should slope into the ground and not onto the plant. In this way, the roots of the plants can be kept cool and moist. The plants must also have shade: dwarf rhododendrons and the lesser Japanese maples are useful in this respect.

At the Royal Botanic Garden, Edinburgh, the Petiolarids are grown in gently sloping beds between bushes which are beneath taller-growing, deciduous trees. The beds are heavily mulched, various substances having been used for this purpose: peat, sawdust (well-rotted, with some bone meal sprinkled on) and spent hops. During summer the beds are kept well hosed, and overhead protection against winter rains is provided by the use of frame lights.

One difficulty with the Petiolarids, and one of their more charming features, is that they flower during the winter months, some of them in the very depths of winter. They therefore need protection against the elements during the winter months. Hence the use of frame-lights, cloches, or variously balanced pieces of glass in the small garden. At this time of year they also benefit, on a dry day, from an occasional watering with liquid Slugit.

Another way of providing winter protection is to lift open ground plants, pot them and take them into the alpine house. The beauty of the individual flowers can then be better seen. But this treatment also creates difficulties, the chief of which is the rotting of leaves caused by the development of moulds.

With potted plants, the soil surface should be given quite a deep covering of granite chips, and this layer taken into the neck of the plant. A dusting of Captan can be applied to the surface of the chips to deter the growth of moulds.

If any leaves are seen to be diseased, then these should be stripped out using a pair of tweezers. Maximum ventilation should be given to plants in the Alpine House, and on fine days the pots can be put in a box and taken outside for a few hours to get increased ventilation. The leaves of potted plants should not be wetted in winter.

13 THE SECTION MALACOIDES

THE SPECIES OF THE SECTION

P. aromatica	*P. effusa*	*P. malacoides*
P. cavaleriei	*P. forbesii*	*P. pellucida*
P. duclouxii	*P. interjacens*	

The species of this section grow in two distinct habitats. *Pp. aromatica, cavaleriei, duclouxii, interjacens* and *pellucida* grow in rock fissures, in shady hollows and in the entrance to caves, usually on limestone. These species are dwarf in stature, with rather delicate foliage. The rootstock is rhizomatous and the species are perennial, with the exception of *P. pellucida*, which is a delicate annual. The above species have only rarely been found and were never in cultivation.

The remaining species, *Pp. effusa, forbesii* and the type species *malacoides*, come from a totally different habitat. They are plants of marshy ground, the fringe of irrigation channels, and rice fields. The species of this group are much more vigorous than the rock-species of the section. *Pp. forbesii* and *malacoides* are annuals.

The distribution of the section is south-western Szechuan, Yunnan, Kweichow and the frontier districts of north-eastern Burma. The various species grow at an average elevation of about 2000 m., and are not hardy in Britain.

P. effusa was grown at Edinburgh during the years 1924–30, but seed was rarely set and the species was subsequently lost. *P. forbesii* was occasionally grown as a greenhouse plant and may still persist. *P. malacoides* is by far the most vigorous member of the section. It has received considerable attention on the part of the horticulturalists and innumerable varieties and forms have been offered as house plants.

[72]

14 THE SECTION GRANDIS

This monotypic Caucasian section is omitted.

15 THE SECTION BULLATAE

The Species of the Section

P. bracteata *P. forrestii *P. redolens
P. bullata P. henrici *P. rockii
*P. dubernardiana

KEY

1 Flowers yellow 2
  Flowers pink, rose, lilac or white 3
2 Leaves tapered at base. Scape 15 cm. or more.
    Flowers 10–25 *forrestii*
  Leaves cordate or square-cut at base. Scape less
    than 15 cm. Flowers 1–5 *rockii*
3 Scape less than 5 cm. Flowers 2–5 *dubernardiana*
  Scape 10–20 cm. Flowers 5–12 *redolens*

The species of this section are woody, evergreen perennials with crinkly (bullate) leaves. This last is not a diagnostic feature: it also occurs in some species of other sections.

Most of the species of this section are farinose. They are rock primulas (often on limestone) and were formerly grouped with the Revolute Fariniferae of similar habitat, such as the section Dryadifolia. But cytological investigation has shown that the species of this section are related to the sections Malacoides and Cortusoides (farinose and efarinose respectively). The species of this section are therefore probably similarly placed to the section Obconica, which has both farinose and efarinose affinities.

The distribution of the section is south-western Szechuan, north-western Yunnan and the adjacent parts of east and south-eastern Tibet.

P. dubernardiana Forrest

P. dubernardiana is a dwarf, shrubby cushion plant with a long, thick, woody rhizome. The rhizome is thickly covered with old

[73]

leaf-scars and, towards the apex, with the twisted, withered remains of dead leaves: a difficult plant to keep alive on anything other than a limestone cliff.

The leaves, including the petiole, are 3–12 cm. long and the lamina 1.5–7 cm. long and 0.5–2 cm. broad, narrowly oblong to oblanceolate, rounded at the apex and tapering rapidly at the base. The margin is entire or crenate and often recurved. The underside of the leaves is covered with copious yellow farina.

The scape is more or less absent and hidden among the leaves, or up to 5 cm., and carries an umbel of 2–5 flowers. The pedicels are 15–25 mm. long.

The corolla is rose with a yellow or orange eye, or white flushed rose. There is no annulus. The corolla-tube is 1–2 cm. long, up to twice the length of the calyx. The limb is 1.5–2 cm. diameter with broadly obcordate lobes deeply notched at the apex.

P. dubernardiana was first found by Forrest, in the valley of the Mekong in north-western Yunnan, in 1904. It grows at about 3000 m. on the dry shady ledges of limestone cliffs, forming dense cushions.

In 1919, flowering plants were raised from seed collected by Forrest in 1917. The species was still in cultivation in 1946, but whether it still persists is not known.

P. forrestii Balf.f. (Plate XIV)

This species is a more or less dwarf-shrubby plant with a long, thick, woody rhizome. The rhizome is covered with numerous leaf-scars and, towards the apex, with the withered remains of old leaves.

The leaves, including the petiole, are 6–20 cm. long and the lamina 3–8 cm. long and 2–5 cm. broad, but larger in fruit, ovate-elliptic or elliptic-oblong to oblong, rounded at the apex, rounded to cuneate at the base, round to sharp-toothed at the margin, green-crinkly above, at first white or pale-yellowish farinose below, but later efarinose.

The scape is 15–(rarely) 90 cm. tall, and carries an umbel of 10–25 flowers. The pedicels are 1–4 cm. long.

The corolla is a deep golden yellow with an orange eye. There is no annulus. The corolla-tube is 1–3 cm. long, 1–2 times the calyx, and the limb is 1.5–2.5 cm. diameter with broadly obovate to subrotundate lobes deeply notched at the apex.

P. forrestii was discovered by Forrest in 1906 on the Lichiang Range in Yunnan, growing 'in the dry shady crevices of limestone cliffs' and seed collected by Forrest in 1906 was flowered in 1909. In the wild it forms a massive, woody rhizome which may persist for fifty years. In Britain it is usually grown in the alpine house. Seed was offered by the Society in 1972 and again in 1975 – viable seed on both occasions.

P. redolens Balf.f. et Ward

This is a robust dwarf shrubby, fragrant plant with a short, stout, woody rhizome which is covered at the apex with the withered remains of old leaves.

The leaves, including the petiole, are 6–12 cm. long. The lamina is 4–6 cm. long and 1.5–3 cm. broad, elliptic to oblong and clearly differentiated from the petiole, rounded at the apex and tapering at the base, round-toothed at the margin, green-crinkled above and sometimes with a little farina below.

The scape is 10–20 cm. tall and carries a single umbel of 5–12 flowers. The pedicels are 1.5–2 cm. long.

The corolla is pale purplish-pink or lilac-pink fading to blush-cream, with a yellow eye, or white ageing to pale violet. There is no annulus. The corolla-tube is 1–1.5 cm. long, twice the length of the calyx, and the limb is 1.5–2.5 cm. diameter with obcordate to rounded lobes deeply notched at the apex.

P. redolens was discovered as a single clump in fruit, on dry cliffs in the Mekong gorge, in 1911. Plants raised from this seed had flowered in a number of gardens in Britain by 1915, but the species rarely set seed in cultivation. *P. redolens* has therefore tended to decline in cultivation and it is doubtful if any still remains.

P. rockii W.W.Sm.

This is a small shrubby plant with a long, thin, woody rhizome which is covered with the remains of old leaves.

The leaves, including the petiole, are 2–10 cm. long. The efarinose lamina is 1–3 cm. long and 1–2.5 cm. broad, broadly ovate, rounded or obtuse at the apex, cordate to square-cut at the base, shallowly lobed at the margin, the lobes bluntly-toothed.

The scape is absent or 5–15 mm. long and carries an umbel of 1–5 flowers. The pedicels are 2–4 cm. long.

The corolla is yellow. There is no annulus. The corolla-tube is 1–2 cm. long, 1–2 times the calyx, and the limb is 1.5–2 cm. diameter with obcordate to rounded lobes shallowly notched at the apex.

This dwarf species was discovered by Dr. Rock in 1928 in the mountains of south-western Szechuan, where it grew on limestone cliffs and boulders at an elevation of 3000–4425 m. The species has

[75]

since been found on a number of occasions, but always in the same area. Plants were raised from Rock's seed and flowered in Britain in 1935, but it is doubtful if the species still remains in cultivation.

16 THE SECTION DRYADIFOLIA

The Species of the Section

P. chlorodryas	*P. jonardunii*	*P. triloba*
P. dryadifolia	*P. mystrophylla*	*P. tsongpenii*

The type species of the section is *P. dryadifolia*. This plant was first collected in 1884 by the Abbé Delavay on the Lichiang Range in Yunnan. The following year it was described by Franchet, who commented on the very similar appearance of *Dryas octopetala*.

The species appears to have been collected next by Forrest in 1905, and he subsequently acquired seed from which plants were raised and flowered at Edinburgh in about 1913–14. Balfour (3) describes *P. dryadifolia* as 'one of the finest of Primulas'.

The characteristics of species of the section include the development of woody rhizomes clothed with dead, withered leaves, as with the sections Bullatae and Auricula, and the cushion-like form of growth; the crowded, evergreen leaves, ovate to orbicular in profile, with diminutive round-toothed margins which are rolled backwards; and the more or less sessile flowers which are as large as the leaves. The corolla is annulate, with lobes shallowly notched at the apex.

The distribution of the section includes the western ranges of Yunnan and Szechuan, the Bhutan and Assam Himalaya, and adjoining parts of south and south-eastern Tibet and north-western Burma. The species of this section therefore occupy much the same area as the subsection Bella of Minutissimae, to which the present section is closely related.

The species of the section are all high alpines, with an altitudinal range of 4000–5500 m. Forrest describes *P. dryadifolia* as growing abundantly on great limestone screes to 5500 m. The species also appear to grow equally on gneiss and granite – near to the glaciers and perpetual snow according to Delavay.

The only two species of the section which appear to have been in cultivation are *Pp. chlorodryas* and *dryadifolia*.† *P. chlorodryas* was

† *P. jonardunii* has also been in cultivation (Ed.)

flowered from seed in about 1934 but there are obvious difficulties in maintaining the woody rhizomic development and the species did not long survive. The flowering of *P. dryadifolia* has already been discussed. This species appears to have persisted until about 1938.

17 THE SECTION MINUTISSIMAE

THE SPECIES OF THE SECTION

SUBSECTION EU-MINUTISSIMAE

P. annulata	*P. muscoides*	*P. stirtoniana*
P. candicans	*P. praetermissa*	*P. subularia*
P. flagellaris	*P. reptans*	*P. tenella*
P. geraldinae	*P. rhodochroa*	*P. tenuiloba*
P. glandulifera	*P. rimicola*	*P. waddellii*
P. heydei	*P. rubicunda*	*P. walshii*
P. minutissima	*P. spathulifolia*	

SUBSECTION BELLA

P. barbatula	*P. moschophora*	**P. primulina*
P. bella	*P. occlusa*	

This section comprises a somewhat heterogeneous collection of dwarf species, which are almost invariably single flowered. The corolla-tube is 2–3 times the length of the calyx and each corolla-lobe is divided to the throat or at least deeply notched at the apex.

Neither are the species all alike in habit. Some show a mat-like growth, with both leaves and flowers scarcely rising above the level of the soil, but the majority of species of the section form neat rosettes with the flowers more or less above the leaves.

There is no vigorous rhizomic development as with species of the section Bullatae, indicative of plants attaining great age. In some species this is replaced by the development of stoloniferous growth, which is an annual feature.

White or yellow farina occurs freely on about half the species. While there appears to be no morphological significance in the presence of the farina, it provides a convenient method of grouping the Section, and those Minutissimae species which come closest to Farinosae are among the farinose members of Minutissimae.

The leaves of species of the section also show a high degree of specific divergence in both shape and size. The margin may be finely

[77]

or coarsely serrate to dentate, or more or less entire. The petiole may be well developed and distinct, or almost absent.

The scape is usually short or almost absent, but is well developed in a few species. The flower is usually solitary, but in *Pp. glandulifera, primulina* and *heydei* the flowers are more numerous.

With species of the subsection Eu-Minutissimae, the throat of the corolla is without hairs or with a few scattered hairs only, whereas species of the subsection Bella have the throat closed with a mop of loose hairs.

With the exception of three species only, the section is confined to the Himalaya – some species favour the western Himalaya and some the eastern Himalaya, but no single species grows throughout the whole length of the Himalayan range. The exceptions are *P. walshii*, an eastern Himalayan plant which is also found in Szechuan and *Pp. bella* and *moschophora* which are found in south-western China and the adjacent parts of Tibet and Burma.

Few species of the section have been in cultivation, and none are likely to prove amenable to cultivation. They are high alpines, growing in the earthy grit of steep slopes and watered by the frequently swirling mists and rain. Such conditions are difficult to emulate in cultivation: the few plants brought back have grown out of character and been but short-lived.

P. primulina Wall. (Illustration p. 21)

P. primulina, which used to be *P. pusilla*, is the sole member of the section likely to be seen in cultivation and even this species must be diligently sought, for it is not normally offered by nurserymen.

The many leaves of plants of this species grow in a close rosette, with the remains of older leaves beneath. The leaves, including the petiole, are 1–3 cm. long and 3–5 mm. broad, spathulate or oblanceolate, coarsely and deeply toothed, rounded in profile at the apex but with projecting teeth, and at the base gradually narrowed into the winged petiole which may equal the blade, but is usually much shorter. Both surfaces of the lamina are efarinose and there are rather long, soft hairs above.

The scape, which does not exceed 8 cm., carries 2–4 flowers which are sessile or nearly so, or occasionally there is only a single flower.

The corolla is purple or violet, rarely white. The corolla-tube is short, up to 5 mm. and scarcely exceeding the calyx, densely covered outside with soft, short hairs or slightly farinose and the throat is blocked by a dense tuft of white hairs. The limb is 8–10 mm.

[78]

diameter, with obcordate lobes shallowly notched at the apex. Both pin-eyed and thrum-eyed forms occur.

P. primulina is a common plant of the eastern Himalaya and is found throughout Nepal, Sikkim and Bhutan, where it grows at an elevation of 4000–5000 m. It has also been found in southern and south-eastern Tibet.

18 THE SECTION SOULIEI

THE SPECIES OF THE SECTION

P. aliciae	*P. longipinnatifida*	*P. souliei*
P. humilis	*P. rupicola*	*P. xanthopa*
P. incisa		

The species of this section are perennials with a rather short, stout rootstock or long, thin rhizome which is usually clothed, near the base, with the remains of old leaves. It is a section of closely related species which also suggests proximity to certain members of the section Farinosae, particularly *Pp. yunnanensis* and *gemmifera*. But the habit is different, and other features of the section Souliei which distinguish its members from Farinosae include the long and distinct petioles, the thin texture of the membranous leaves which are somewhat rough to the touch on the upper surface and farinose below, with white or pale yellow farina. The margin of the leaves may be more or less entire (*P. souliei*) to markedly serrate and deeply and pinnatifidly cut (*Pp. incisa* and *longipinnatifida*).

These are generally dwarf plants, with the scape not exceeding about 12 cm., but 5–45 cm. in *P. incisa*. The inflorescence is a loose to tight umbel.

The distribution of the section includes the western provinces of China, eastern Tibet and the eastern Himalaya.

Most species of the section are high Alpines growing on granite rocks, on rock-ledges and stony alpine pasture at 4000–5000 m.; but *Pp. incisa* and *longipinnatifida* also descend into the sub-alpine woods and forests at about 3000 m.

The species of the section are rather intractable in cultivation. *Pp. aliciae, incisa, rupicola* and *xanthopa* have been in somewhat transient cultivation, although *P. incisa*, which is much the most

[79]

vigorous species of the section, was kept for nearly twenty five years until finally lost in 1939. *Pp. humilis, longipinnatifida* and *souliei* have never been in cultivation.

19 THE SECTION FARINOSAE

THE ASIATIC SPECIES OF THE SECTION

SUBSECTION EU-FARINOSAE

*P. algida	P. erratica	P. knuthiana
P. assamica	P. farinosa	P. matsumurae
P. baldschuanica	*P. fauriae	*P. modesta
P. blandula	P. fernaldiana	P. schlagintweitiana
P. borealis	P. fistulosa	*P. sharmae
*P. caldaria	P. flava	P. sorachiana
*P. capitellata	*P. hazarica	*P. stenocalyx
P. clutterbuckii	*P. intermedia	*P. tayloriana
P. concinna	*P. jaffreyana	*P. yuparensis

SUBSECTION GEMMIFERA

P. conspersa	*P. gemmifera	P. laciniata

SUBSECTION PULCHELLA

P. prattii	*P. pulchella	*P. pulchelloides

SUBSECTION YUNNANENSIS

P. fragilis	P. membranifolia	P. umbrella
P. kialensis	P. nutantiflora	P. yunnanensis

SUBSECTION INAYATII

*P. inayatii

SUBSECTION GLABRA

P. genestieriana	P. kongboensis	P. pseudoglabra
*P. glabra		

SUBSECTION SIBIRICA

P. fasciculata	P. pumilio	*P. tibetica
*P. involucrata	*P. sibirica	P. urticifolia
P: oxygraphidifolia	P. tanupoda	*P. yargongensis

*P. auriculata	P. fedschenkoi	*P. rosea
*P. clarkei	P. flexuosa	P. sertulum
P. efarinosa	P. macrocarpa	*P. warshenewskiana
P. elliptica	P. obsessa	

KEYS

KEY TO SUBSECTIONS

1 Leaves (at least the undersides) farinose   2
  Leaves efarinose†   6

2 Corolla-tube less than twice as long as calyx   Eu-Farinosae
  Corolla-tube more than twice as long as calyx   3

3 Farina white   Eu-Farinosae
  Farina yellow   4

4 Seed-capsule longer than calyx   Pulchella (& P. modesta)

  Seed-capsule shorter than calyx   5

5 Leaves 2–5 times longer than broad (oblong to elliptic)   Yunnanensis (& P. sharmae)
  Leaves 6–12 times longer than broad (strap-shaped)   Inayatii

6 Scape and calyx white-farinose. Basal buds present   Gemmifera
  Scape and calyx efarinose   7

7 Pedicels usually shorter than calyx   8
  Pedicels longer than calyx   9

8 Corolla-tube, capsule and calyx approx. the same length. Flower very small   Glabra
  Corolla-tube 2–3 times capsule and calyx. Flowers conspicuous   (P. auriculata)

9 Leaves more or less entire. Capsule longer than calyx   Sibirica
  Leaves clearly toothed. Capsule = calyx   Auriculatae

† Some species of Eu-Farinosae, viz. Pp. farinosa, algida, intermedia, stenocalyx, hazarica and jaffreyana, are, or may be, efarinose.

KEY TO SUBSECTION EU-FARINOSAE

1 Flowers clearly annulate   2
  Flowers not, or doubtfully, annulate   6

2 Flowers white; petals shallowly notched.
  Stoloniferous   *caldaria*
  Flowers pink to violet; petals deeply lobed. No
    stolons   3
3 Corolla-tube 2–3 times longer than calyx   4
  Corolla-tube less than 2 times longer than calyx   5
4 Leaves sharply and irregularly toothed   *jaffreyana*
  Leaves entire or with minute round teeth   *stenocalyx*
5 Pedicels 6 mm. or less   *algida*
  Pedicels 7 mm. or more   *fauriae*
6 Under-surface of leaves efarinose   *intermedia*
  Under-surface of leaves farinose, at least
    transiently   7
7 Farina white   8
  Farina yellow or cream   10
8 Pedicels 0–1 mm.   *capitellata*
  Pedicels 7 mm. or more   9
9 Petals deeply lobed   *hazarica*
  Petals shallowly notched   *yuparensis*
10 Farina cream. Flower violet with whitish-yellow
    eye   *tayloriana*
  Farina yellow. Flower pink to mauve, no distinct
    eye   11
11 Leaf-stalks as long as blade, or longer; pedicels
    10 mm. or more   *modesta*
  Leaf-stalks one-third as long as blade, or less;
    pedicels 7 mm. or less   *sharmae*

KEY TO SUBSECTION SIBIRICA

1 Flowers white; petals shallowly notched   *involucrata*
  Flowers pink, mauve or purple; petals deeply
    lobed   2
2 Scape usually very short; corolla-tube less than
    10 mm.; limb less than 10 mm. diameter   *tibetica*
  Scape usually 10 cm. or more; corolla-tube 10 mm.
    or more; limb more than 10 mm. diameter   3
3 Bracts with teeth 4–7 mm. long   *yargongensis*
  Bracts without such teeth   *sibirica*

KEY TO SUBSECTION AURICULATAE

1 Umbels dense; flowers sessile or nearly so   *auriculata*
  Umbels loose; pedicels distinct   2

[82]

2 Leaf-blades as broad (or nearly) as long, not
tapered into petiole; petiole 2–3 times as long
as blade *clarkei*
Leaf-blades much longer than broad, tapered into
petiole; petiole not longer than blade 3
3 Upright plant. Scape usually more than 5 cm. at
flowering *rosea*
Dwarf spreading plant. Scape usually less than
5 cm. at flowering *warshenewskiana*

With about eighty species, the section Farinosae is the largest in the
genus, and is divided into a number of subsections of which the
major division is the Eu-Farinosae, based on *P. farinosa*. The leaves
of species of this subsection show a marked development of farina on
the leaves, but a few species are efarinose or nearly so. In other subsections the leaves are efarinose or, if farinose, have yellow farina.
The subsection Eu-Farinosae has the widest distribution. Most of the
species are Asiatic, but several species occur in northern Europe and
a few in North America and the extreme south of South America.

Closely related to the Eu-Farinosae is the subsection Sibirica, the
species of which are efarinose and the leaves more or less entire. The
distribution is mainly Asiatic, with a few species in arctic America
and northern Europe.

Of the remaining subsections, five are wholly Asiatic and the sixth,
Auriculatae, crosses the European boundary in the Caucasus with
the species *Pp. auriculata* and *luteola*.

Also near to Eu-Farinosae are the subsections Pulchella and
Gemmifera. Both are confined to the western provinces of China and
the adjacent Tibetan marches. In both subsections the capsule tends
to exceed the calyx. Species of the subsection Pulchella have rather
fleshy leaves the underside of which is usually densely covered with
yellow farina. The subsection Gemmifera contains species with
efarinose leaves, but the scape and calyx are strongly white-farinose.

The other four subsections all have a capsule enclosed by the calyx,
or not exceeding the calyx. The subsections Yunnanensis and Glabra
have closer affinities with Eu-Farinosae than with Auriculatae (which
was formerly a separate section). The distinguishing features of
Yunnanensis include the short capsule and the yellow farina which
covers the underside of the leaves. This subsection is confined to the
western provinces of China and to northern Burma. The subsection
Glabra comprises dwarf species which are efarinose or nearly so,

with a short capsule. Distribution includes the eastern Himalaya, Tibet and western China.

The subsection Inayatii consists of a single species which is found in the north-western Himalaya. In its capsule and seed it is rather similar to P. *rosea* (subsection Auriculatae) but the underside of the strap-shaped leaves is heavily coated with yellow farina. The subsection Inayatii therefore lies somewhere between the subsections Eu-Farinosae and Auriculatae.

Most members of the section are of average stature or less, and some are even dwarf. With many species, particularly the Eu-Farinosae, the rhizome is but poorly developed. This may be one reason for the comparatively short life of plants, particularly of this subsection, in cultivation. The species more amenable to cultivation include certain members of the subsections Auriculatae, Inayatii and Sibirica which have stronger rhizomes.

About half the species of the section have probably been in cultivation at one time or other. But even in the wild these are not long-lived plants. To maintain a succession the problem is, therefore, to obtain viable seed – and this is not always produced in cultivation.

One final general point of interest concerning species of the section Farinosae is that they do not appear to hybridize in the wild, and hardly ever in cultivation.

THE SUBSECTION EU-FARINOSAE

P. algida Adam (Plate XV)

This perennial species forms a short thick rhizome which, at flowering time, has no basal scales.

The leaves, including the petiole, are 3–4 cm. long and 1–1.5 cm. broad, oblong to oblanceolate, more or less rounded at the apex and tapering at the base into the broadly winged petiole, which is usually short but rarely equals half the blade. The margin is finely toothed, or more rarely entire, efarinose above and clothed below with yellow or white farina, or efarinose in var. *sibirica*.

The scape is 3–20 cm. tall in flower, but may be nearly twice this length in fruit, and carries a more or less head-like umbel of 3–12 flowers. The pedicels are 1.5–6 mm. long, but lengthen in fruit to 15 mm.

The limb of the annulate corolla is violet or rarely white, and the corolla-tube is yellowish or white. The tube is 5–10 mm. long, equalling the calyx, and the limb is 8–15 mm. diameter with obcordate lobes deeply notched at the apex.

[84]

P. algida grows in moist alpine and subalpine meadows at 3000–5000 m. and flowers in May and June. Its distribution extends from the Caucasus to Turkestan, northern Iran and the Altai. The leaves may be efarinose or heavily coated with farina below. Turkeviez states that this feature varies with geographical grouping: in the Altai efarinose forms alone occur, in Samarkand only the farinose, while in Turkestan both forms occur growing together, with the farinose species more frequent.

Like many species of this section, *P. algida* is short-lived in cultivation and has to be raised from seed, preferably imported wild seed. In 1975 the Scottish Rock Garden Club offered viable seed, with no indication of its origin.

P. caldaria W.W.Sm. et Forrest

This species is a perennial with a stout rhizome from which grow leafy stolons 5–20 cm. long.

The leaves are 3.5–25 cm. long and 6–35 mm. broad, oblanceolate, rounded or obtuse at the apex, at the base gradually tapered into the winged petiole which equals or exceeds the blade. The margin of the lamina is bluntly and irregularly toothed in the upper half. The upper surface of the lamina is glabrous and the lower surface covered with white farina which rapidly disappears.

The scape is 10–50 cm. tall, flexuous, and carries an umbel of 20–30 flowers. The pedicels are 8–16 mm. long, slender and erect.

The annulate corolla is white. The corolla-tube is 6–7 mm. long, about 1½ times the calyx. The limb is 5–7 mm. diameter, with obcordate lobes shallowly notched at the apex.

This species was first collected by Forrest on the Mekong-Salween divide, growing at 2700–3000 m. in the lush herbage near hot water springs 'as its specific name is intended to imply' (5). The species was subsequently found by Ward in the bed of the Mekong river where streams come down from limestone country above.

P. caldaria has rarely been found, but both Forrest and Ward obtained seed from which plants were flowered at Edinburgh in 1925. It is doubtful however if the species still remains in cultivation.

P. capitellata Boiss.

P. capitellata is a species which varies much in stature. It is perennial with a short, stout rhizome the upper part of which, at flowering time, has a few persistent scales.

The leaves are 1.5–10 cm. long and 5–15 mm. broad, oblong-lanceolate to spathulate, obtuse, rounded or even acute at the apex, at the base narrowed into a short, broadly sheathing petiole. The lamina is more or less entire to shallowly or markedly toothed at the margin, glabrous above and covered below with a white farina which quickly disappears.

The scape is 2.5–25 cm. tall and carries a dense head of 5–10 flowers. The pedicels are absent or very short, rarely exceeding 1 mm. The faintly annulate corolla is rose. The corolla-tube is 6–8 mm. long, about 2 times the calyx. The limb is 6–7 mm. diameter, with obcordate lobes shallowly notched at the apex.

P. capitellata was first collected in southern Iran in 1845 by Kotschy. It was found growing in moist alpine meadows at 3000–3600 m. and appears to be not uncommon in the mountains of Iran. It has also been recorded from Afghanistan.

The species was in flower at Edinburgh in 1925, since when it has been flowered, somewhat intermittently, on a number of occasions. Like many species of the section, it is readily lost.

P. farinosa Linn.

P. farinosa is the most widely distributed species of the genus. It occurs throughout northern Asia to the shores of the north Pacific and southwards to the Altai and Tien Shan. But it does not occur in the Himalaya or Japan and in China only in Kansu. Neither does it occur in either North or South America, but it grows in all the countries of Europe except Greece and Ireland. It is to be described in the companion publication on European and other primulas.

P. fauriae Franch.

P. fauriae is a dwarf farinose plant with a short, slender rhizome.

The leaves, including the petiole, are 2–6 cm. long. The lamina is 8–12 mm. long and almost as broad, ovate or rhomboid or suborbicular, rounded or obtuse at the apex, and more or less truncate at the base or tapering into the narrowly winged petiole, which is much longer than the lamina. The margin is more or less entire or finely crenulate and the blade is yellow-farinose below.

The scape is 3–12 cm. high and carries an umbel of 2–9 flowers. The pedicels are 7–30 mm. long, unequal and stiffly erect.

The markedly annulate corolla is rose or purplish. The corolla-tube is 3–5 mm. long, equalling the calyx or a little longer. The limb is 7–8 mm. diameter with spreading, narrowly obcordate lobes deeply divided from the apex.

This is a Japanese species, first found by Faurie on Mt. Shichinoke in 1886. It is kept in cultivation, somewhat transiently, by raising plants from seed, which appears to have been last available from the Society's exchange in 1973.

P. hazarica Duthie

P. hazarica is a farinose plant with a well-developed rhizome which, at flowering time, has no basal scales.

The leaves, including the petiole, are 1.5–10 cm. long and 0.75–2 cm. broad. The lamina is spathulate to obovate, rounded at the apex, and tapers into the winged petiole, which may be very short or more or less equal to the lamina. The margin is crenulate, or sharp toothed, and the lamina is thinly covered below with white farina which quickly disappears.

The scape is 1–10 cm. tall and carries an umbel of 2–12 flowers. There are 1–6 scapes to each plant. The pedicels are 1–2.5 cm. long.

The corolla is mauve-purple with a pale yellow tube. There is no annulus. The corolla-tube is 1–2 cm. long, 2–3 times the calyx. The limb is 1–1.5 cm. diameter with obcordate lobes deeply notched at the apex.

This species is found in Hazara and Kashmir in the north-west Himalaya, where it grows at 3300–4700 m. in very moist rock fissures. Seed collected in 1935 by Ludlow & Sherriff gave rise to plants which flowered in Edinburgh in 1936–8, but the species does not now appear to be in cultivation.

P. intermedia Sims

P. intermedia is a perennial with a short rootstock. For the section, it is a comparatively tall species.

The leaves, including the petiole, are 3–13 cm. long and 0.7–2 cm. broad, oblong-obovate or oblong to lanceolate, rounded at the apex and, at the base, tapering into the winged petiole which is usually very short, but may equal half the blade. The margin of the lamina is entire or wavy or minutely round-toothed, and the surface of the lamina is efarinose and slightly glaucous.

The scape is 10–30 cm. tall in flower, but elongates in fruit to 35–60 cm. It is stout and erect and carries a single umbel of from 10 to about 100 flowers. The pedicels are 3–10 mm. long at early flowering, erect, and 6–30 mm. in fruit.

The corolla is rosy-violet and more or less annulate. The corolla-

[87]

tube is 5–7 mm. long and the flat limb 6–10 mm. diameter with obcordate lobes deeply notched at the apex.

The distribution of the species extends from western Siberia to the Alatau, Tien Shan and Altai in Turkestan. It grows in the moist meadows of river valleys, particularly in steppe country.

P. intermedia was so named as being intermediate in character between *P. farinosa* and *P. auriculata*. It was introduced into Britain in 1809, but was always rare, and is probably now no longer in cultivation.

P. jaffreyana King

This is a farinose plant with a short rhizome which is without basal scales.

The leaves, including the petiole, are 2–20 cm. long and 1–4 cm. broad, but up to 25 cm. long in autumn. In profile, the lamina is obovate-elliptic to oblong or oblanceolate, rounded or obtuse at the apex, at the base tapering into the winged petiole which is usually shorter than the blade, but is sometimes more or less equal to it. The margin is usually sharply toothed. The lamina is membranous and the underside is covered with white farina which disappears with time.

The scape is 5–25 cm. tall and carries an umbel of 3–13 flowers. The pedicels are 0.5–2.5 cm. long.

The annulate corolla is mauve to pale blue-violet with a green or yellow eye. The corolla-tube is 1.5–2.5 cm. long, about 2 times the calyx. The limb is 1–2 cm. diameter with obcordate lobes deeply notched at the apex.

P. jaffreyana is found in the Chumbi valley and that part of south Tibet to the immediate north of Bhutan. It grows at 3300–5300 m. on dry sandy banks either in the open or in mixed forest. The species was introduced into cultivation by means of the winter-resting buds which, in the dry conditions of the natural habitat, are all very small. Seed collected by Sherriff in 1938 flowered at Edinburgh in 1939, and the species may still persist.

P. modesta Bisset et Moore (Illustration p. 21)

This species is a dwarf, farinose plant with a short, stout rootstock. There are usually neither basal scales nor the persistent remains of withered leaves. The winter-resting condition is a tight, contracted bud.

Plate XXI *P. bellidifolia* (p. 103)   *Photo: Royal Botanic Garden, Edinburgh*

Plate **XXII** *P. chasmophila* (p. 109) *Photo: Royal Botanic Garden, Edinburgh*

Plate **XXIII** *P. sikkimensis* (p. 134) *Photo: Royal Botanic Garden, Edinburgh*

Plate XXIV *P. eburnea* (p. 109)   *Photo: Royal Botanic Garden, Edinburgh*

Plate XXV *P. reidii*, var. *williamsii* (p. 111)   *Photo: R. C. Elliott*

Plate XXVI *P. melanops* (p. 121) *Photo: Royal Botanic Garden, Edinburgh*

Plate XXVII *P. sinopurpurea* (p. 124)   *Photo: W. Schacht*

Plate XXVIII *P. anisodora* (p. 137)   *Photo: Royal Botanic Garden, Edinburgh*

Plate XXIX *P. helodoxa* (p. 141)   *Photo: Royal Botanic Garden, Edinburgh*

The leaves, including the petiole, are 2–8 cm. long and 1–1.5 cm. broad, oblong-elliptic to spathulate, rounded or obtuse at the apex, and at the base tapering into the winged petiole which equals or exceeds the blade. The lamina is undulate-crenate or bluntly serrate at the margin and thickly covered below with yellow farina.

The scape is 3–14 cm. tall and carries an umbel of 2–15 flowers. The pedicels are 1–2.5 cm. long and more or less erect.

The slightly annulate corolla is pinkish-purple. The corolla-tube is 6–15 mm. long, 2–3 times the calyx. The limb is 1–1.5 cm. diameter with obcordate lobes deeply notched at the apex.

P. modesta is confined to the alpine regions of Japan. It was introduced into cultivation at Edinburgh in 1911 and appears to have proved nearly as amenable as the European *P. frondosa*. It is rarely offered by nurserymen, but viable seed was included in the Society's 1975 exchange list.

P. sharmae H. R. Fletcher

This species was first found in 1931 by Sharma, growing at 4300–5000 m. in Nepal. It is related to *P. schlagintweitiana* from the north-western Himalaya, and these two species appear to replace *P. farinosa* in the western and central Himalaya.

The leaves, including the petiole, are 4–6 cm. long and 8–10 mm. broad, spathulate or oblong-elliptic, acute to obtuse or rounded at the apex, and at the base tapering into the winged petiole which may be short, or equal to a third of the blade. The margin is finely toothed or more or less entire and the underside of the lamina is densely yellow-farinose.

The scape is 5–10 cm. high and carries 3–8 flowers. The pedicels are 2–7 mm. long, lengthening in fruit.

The corolla is mauve-purple or bluish-purple. There is no annulus. The narrow corolla-tube is 1–3 cm. long, 2–3 times the calyx and the limb is 1.5–2 cm. diameter with spreading, broadly obcordate lobes deeply notched at the apex.

The species was flowered at Edinburgh in 1933, but persisted for one or two seasons only.

P. stenocalyx Maxim.

This species produces a stout rhizome which is free from basal scales.

The leaves, which resemble those of the common daisy, grow in

[89]

a compact to loose rosette. Including the petiole, they are 1–5 cm. long, and 0.5–1.5 cm. broad, obovate or oblanceolate to spathulate, rounded at the apex and at the base tapering into the membranously winged petiole, which may be quite short or more or less equal to the lamina. The blade is efarinose above, and may be efarinose or densely white or yellow-farinose below.

The scape is 1–15 cm. tall, with an umbel of 4–16 flowers. The pedicels are 0–4 cm., stiff and erect, or slender and flexuous.

The annulate corolla is bluish-lavender to pink. The corolla-tube is 1–2 cm. long, 2 times the calyx, or much less, and the limb 1.5–2 cm. diameter with spreading, obcordate lobes deeply notched at the apex.

The distribution of the species is confined to Kansu and Szechuan in western China, and eastern Tibet. It grows on dry, crumbling soils on warm sunny banks at 2700–4300 m., or on limestone.

The white-farinose form was in flower at Edinburgh in 1936, but is probably now lost.

P. tayloriana H. R. Fletcher

This dwarf perennial has a short but fairly thick rootstock from which descend the numerous fibrous roots. The winter-dormant form is a farinose resting bud not exceeding 1 cm.

The leaves, including the petiole, are 4–5 cm. long and 2–3 cm. broad, obovate, rounded at the apex and tapering at the base into the petiole which is scarcely differentiated from the lamina. The margin is shallowly sharp- or round-toothed. The lamina is without hairs, efarinose above and covered beneath with cream-coloured farina.

The scape is 4–10 cm. tall, fairly stout for its size, and carries an umbel of 2–8 flowers. The pedicels are 5–15 mm. long. The corolla is violet with a whitish-yellow eye. There is no annulus. The narrowly cylindrical corolla-tube is 1–2 cm. long, 2–3 times the calyx, and the limb is 2–2.5 cm. diameter with obcordate lobes deeply notched at the apex.

This is a species of south-eastern Tibet. It was collected by Ludlow, Sherriff and Elliot in 1947. Its habitat is given as rock faces in a ravine bed at an altitude of 10,000–11,000 ft.

P. tayloriana was grown in cultivation, flowered well and is featured in the pages of the *Bulletin*. But like many species of this section, *P. tayloriana* is not a vigorous plant. There comes a year when seed is not set, or seed does not germinate, and in the absence of fresh seed the species is lost.

P. yuparensis Takeda

This species is a perennial with a short rootstock which is some-times clothed with the withered remains of old leaves.

The leaves are 1.5–3 cm. long and 5–10 mm. broad, oblanceolate or more or less elliptic, rounded or obtuse at the apex, and gradually tapered at the base into the very short winged petiole. The lamina is minutely toothed at the margin, efarinose above and sparingly covered below with white farina which soon disappears.

The scape is 4.5–6 cm. high and bears an umbel of 2–3 flowers. The pedicels are erect and 7–13 mm. long.

The corolla is purple. There is no annulus. The corolla-tube is 12–14 mm. long, 2 times the calyx. The limb is about 1.5 cm. diameter with obcordate lobes shallowly notched at the apex.

This Japanese species was first collected, in 1913, on the summit of Mt. Yuparo, and appears to be quite local in its distribution.

In 1975 seed was offered by the Society's exchange.

THE SUBSECTION GEMMIFERA

The common characteristics of this subsection include the produc-tion of basal buds, the efarinose leaves, the white-farinose in-florescence, a tendency to obliqueness of the corolla, the corolla-tube twice the length of the calyx, and the capsule exceeding the calyx.

The distribution of the subsection includes the three western provinces of China: Kansu, Szechuan and Yunnan, adjacent parts of Tibet and the north Chinese province of Shansi.

P. gemmifera Batalin (Illustration p. 22)

This species is a short-lived perennial which produces one or more basal buds. One of these buds persists in contact with the parent-stock, which it subsequently replaces.

The leaves, including the petiole, are 1–7 cm. long and 5–20 mm. broad, oblong to broadly spathulate, obtuse or rounded at the apex and at the base gradually tapered into the winged petiole which may be short, but often equals the blade and may be twice as long. The lamina is efarinose and rather remotely and irregularly toothed at the margin.

The scape is 8–30 cm. high, sturdy, and carries an umbel of 3–10 flowers. The pedicels are 6–35 mm. long and erect, but pendant near the apex.

The annulate corolla is soft pink to purple. The corolla-tube is

1-3 cm. long, 2-3 times the calyx. The limb is 1.5-2.5 cm. diameter, often slightly oblique or reflexed, with obcordate lobes shallowly notched at the apex.

The distribution of the species includes western Kansu and Szechuan and eastern Tibet. It grows at 3000-4000 m. in peaty turf and in the silty margin of streams.

P. gemmifera was brought into cultivation from seed collected by Farrer in 1914 and may still persist.

var. *zambalensis* W.W.Sm. et Fletcher

This variety occurs in the southern part of the area of distribution of the species. It is of stouter habit and has larger, exannulate flowers. It grows in moist, stony pastures of 3300-5000 m. and is of easier cultivation than the species.

THE SUBSECTION PULCHELLA

The species of this subsection come from south-western China. In habit and foliage, they are somewhat similar to both Farinosae and the Nivalids; but in floral structure they are dissimilar to the Nivalids and the capsule and seed confirm their affinities with the Farinosae.

P. pulchella, which is described as 'a beautiful species' (5) was first discovered in 1884 by the Abbé Delavay in Yunnan. It grows in alpine meadows in limestone country at an elevation of 2000-4500 m. and appears to be fairly common.

The species was brought into cultivation in 1908 from seed sent back by Forrest. It remained in cultivation somewhat transiently until 1934, but is probably now lost. The corolla is pale lavender to deep purplish blue with a yellowish-green eye.

P. pulchelloides may be regarded as a dwarf form of *P. pulchella*, but it retains its features in cultivation. The corolla is pale lavender-blue with a yellow eye.

THE SUBSECTION INAYATII

P. inayatii Duthie (Plate XVII)

This species is the sole member of the subsection. It is a perennial with a thick, elongated rootstock which is without basal scales.

The leaves, including the petiole, are 6-10 cm. long and 1-1.5 cm. broad at flowering time (which is January to March) but in autumn are 10-35 cm. long and 1-3 cm. broad, more or less strap-shaped, rounded or obtuse at the apex, and gradually tapering at the base into the membranously winged reddish-brown petiole, which is equal to

[92]

or longer than half the blade. The margin of the lamina is finely serrate to almost entire and the lower surface is usually covered with sulphur-yellow or sometimes cream-coloured farina.

The reddish scape is 7–12 cm. tall when in flower, but elongates to 25 cm. in fruit, and carries an umbel of 3–15 flowers. The stiff pedicels are 1–1.5 cm. long, lengthening to 3 cm. in fruit.

The corolla is bluish-purple to lilac with a yellow eye. There is no annulus. The corolla-tube is 1–1.5 cm. long, about twice as long as the calyx. The limb is 12–16 mm. diameter, with obcordate lobes shallowly notched at the apex.

This is a species of the north-western Himalaya: Hazara and western Kashmir, where it grows at 1350–1800 m. It is apparently a rather local plant for it has not often been found.

The species appears to be in cultivation, but is rather scarce.

THE SUBSECTION GLABRA

P. glabra Klatt

P. glabra is a dwarf plant with a short stout rhizome which is clothed above with the remains of old leaves and with a few scales.

The leaves grow in a compact rosette and, including the petiole, are 1–3 cm. long and 4–10 mm. broad, spathulate, rounded or obtuse at the apex, at the base tapered into the membranously winged petiole which may be very short or as long as the blade. The lamina is efarinose and serrate or raggedly toothed at the margin with the teeth often recurved.

The scape is 2–11 cm. tall and carries an umbel of 6–9 flowers. The pedicels are 1–5 cm. long.

The annulate corolla is pinkish-purple to bluish-violet or rarely white, with a yellow or orange eye. The corolla-tube is 3–5 mm. long, about equal to the calyx. The limb is 4-7 mm. diameter with obovate lobes deeply notched at the apex.

This species grows at 4000–5000 m. on cliffs and grassy alps. Its distribution includes Nepal, Sikkim, Bhutan and southern Tibet.

In cultivation, *P. glabra* benefits from the protection of a sheet of glass in winter. It is only occasionally available.

THE SUBSECTION SIBIRICA

P. involucrata Wall.

This species is an efarinose perennial with a short rootstock, which is quite free of the withered remains of dead foliage.

[93]

The leaves, including the petiole, are 3–15 cm. long and 0.5–2 cm. broad. The lamina is ovate or oblong, obtuse or rounded at the apex, at the base cuneate or rounded. The margin is entire or remotely and obscurely toothed. The petiole is usually 2–3 times the lamina.

The scape is 10–30 cm. tall and bears an umbel of 2–6 nodding flowers. The pedicels are 1–2 cm. long in flower, but much longer in fruit.

The annulate corolla is white with a yellow eye, or occasionally tinged purple or puce. The corolla-tube is 1–1.5 cm. long, about 2 times the calyx, and the limb is 1.5–2 cm. diameter with obcordate lobes shallowly notched at the apex.

P. involucrata is a widely distributed species occurring throughout most of the Himalayan ranges. It grows at 4000–5000 m. in moist meadows and on the banks of streams.

In cultivation this is a fairly compact plant which does not expand greatly in late summer and autumn, but its liking for moisture must be catered for. It flowers early and may be lifted and potted for better protection of the flowers, but open-ground plants grow much better.

P. sibirica Jacq.

P. sibirica is an efarinose perennial, with a short thin rhizome, which has usually neither basal scales nor the withered remains of dead leaves.

The leaves, including the petiole, are 1–12 cm. long. The blade is 1.5–5 cm. long and 0.5–1.5 cm. broad, ovate or oblong, rounded at the apex, tapering or more abruptly narrowed at the base into the slender, narrowly-winged petiole, which is more or less equal to or 2–4 times the lamina. The margin is entire or obscurely toothed.

The scape is 2–30 cm. tall and carries an umbel of 1–10 flowers. The pedicels are 0.5–4.5 cm. long.

The annulate corolla is pinkish purple or lilac with a yellow eye. The corolla-tube is 1–1.5 cm. long, 1–2 times the calyx, and the limb is 1–2 cm. diameter with obcordate or broadly cuneate lobes deeply notched at the apex.

This species is widely distributed in arctic and central Asia, Kansu, Tibet and the north-western Himalaya. It also extends into arctic Europe and Alaska. It is a species which is rather variable in both habit and stature, as is shown by the considerable number of synonyms and varieties attributed to it.

P. sibirica has been in cultivation for something approaching a hundred and fifty years, but not continuously. *Pp. involucrata* and *yargongensis*, which belong to the same subsection and are therefore closely related to *P. sibirica*, are better doers.

P. tibetica Watt

P. tibetica is a usually dwarf, efarinose plant with a short rhizome devoid of bud scales.

The leaves, including the petiole, are 1–5 cm. long and 2–10 mm. broad, ovate, elliptic or spathulate, obtuse or rounded at the apex, and tapering at the base into the petiole which may be more or less equal to or 2–4 times as long as the lamina. The margin is entire.

There are 1–6 scapes to each plant, sometimes short and hidden by the leaves, or equal to or 2–4 times the leaves. Each scape carries 1–10 flowers. The pedicels are to 7 cm., and usually much longer than the scape.

The annulate corolla is deep rose or purplish-pink with a yellow eye. The corolla tube is 5–10 mm. long, usually a little longer than to twice the calyx. The limb is 7–10 mm. diameter with spreading, obcordate lobes deeply notched at the apex.

The distribution of *P. tibetica* is southern Tibet and, to the south of the main Himalayan range, in a few areas of Nepal, Sikkim and Bhutan, at an altitude of about 5000 m.

The species has been in cultivation on a few occasions since 1916, but always as a result of the importation of fresh seed, for *P. tibetica* is a short-lived plant.

P. yargongensis Petitm.

P. yargongensis is an efarinose perennial with a short thin rhizome, which bears neither basal scales nor the withered remains of old leaves.

The leaves, including the petiole, are 2–12 cm. long. The blade is 1–6 cm. long and 0.5–3 cm. broad, ovate to lanceolate-oblong, rounded or obtuse at the apex, at the base cuneate or tapered into the winged petiole. The margin is entire or finely and irregularly toothed.

The scape is 10–30 cm. high, many to each plant, each scape bearing an umbel of 3–8 flowers. The pedicels are 1–3.5 cm. long in flower, but up to 8 cm. in fruit.

The markedly yellow-annulate corolla is pale mauve to pink or purple, or rarely white, with a yellow throat and eye which is surrounded by a narrow five-lobed area of white. The corolla-tube is

[95]

1–2 cm. long, not exceeding 2 times the calyx. The limb, which is sometimes slightly oblique, is 2–2.5 cm. diameter with obcordate lobes deeply notched at the apex.

The distribution of *P. yargongensis* is south-western Szechuan, north-western Yunnan and south-eastern Tibet. It grows in wet meadows and beside streams at 3000–4500 m.

This species is well established in cultivation, but perhaps not widely grown. It is a compact plant, very floriferous, and sets seed, provided it gets ample moisture during the growing season.

THE SUBSECTION AURICULATAE

P. auriculata Lam.

P. auriculata is a stout perennial with a well-developed rhizome, which is surmounted, towards the apex, by the withered remains of old leaves.

The leaves, including the petiole, are 3–25 cm. long and 0.7–3 cm. broad, elliptic or oblanceolate to oblong-spathulate, rounded or obtuse at the apex, at the base tapering into the winged petiole which is usually very short, but may be more or less equal to the blade. The lamina is crenate-dentate or remotely and finely toothed, or more or less entire at the margin, and efarinose both surfaces.

The scape is usually solitary, 8–60 cm. high and bears a compact umbel of 8–20 flowers. The pedicels are 1–7 mm. long.

The annulate corolla is rose, reddish-purple or bluish, fading almost to white towards the centre, with a yellow eye. The corolla-tube is 1–2 cm. long, 2–3 times the calyx. The limb is 1–2 cm. diameter with broadly obovate to almost round lobes deeply notched at the apex.

The distribution of the species is Iran, Asiatic Turkey and northwards to the Caucasus. *P. auriculata* grows in wet meadows and by the banks of streams at 2000–2700 m., with alpine forms at 3000–4000 m. The species is rarely offered by nurserymen, but viable seed was offered in the Society's exchange in 1975.

P. clarkei Watt (Plate XVI)

This is a tiny species with a few basal scales.

The leaves are somewhat scattered, not forming a rosette. The petiole is 2.5–3.5 cm. long, usually 2–3 times the blade and more or less winged. The blade is 1.2–1.7 cm. long and 1–2 cm. broad, orbicular or reniform or broadly ovate, rounded at the apex, cordate or more or less truncate at the base. The lamina at the margin is finely sharp- or round-toothed and efarinose both surfaces.

The flowers are usually borne singly on pedicels 3–5.5 cm. long, or a scape may develop to 5 cm. with an umbel of 2–6 flowers and pedicels 1–2.5 cm. long.

The corolla is rose-pink. There is no annulus. The corolla-tube is 1–1.5 cm. long, twice as long as the calyx, and the limb is 1.2–1.8 cm. diameter with obovate lobes deeply notched at the apex.

This species appears to be confined to Kashmir in the north-western Himalaya, where it grows at about 2500 m. It is a dwarf species, which lends itself to pan-culture and is frequently seen on the show benches.

P. rosea Royle

P. rosea is an efarinose plant with a short stout rhizome, which is clothed with many, overlapping, broadly ovate bud-scales.

The scape develops before the leaves, and the buds show colour as soon as they push through the earth, before there has been time for development of the scape.

The leaves therefore mature at late flowering or fruiting time when, including the petiole, they are 3–20 cm. long and 0.5–4 cm. broad, ovate-oblong or obovate to oblanceolate, rounded or obtuse at the apex, at the base tapering into the winged petiole which is usually absent at flowering but subsequently lengthens to half the blade and may equal the blade. The margin is fairly regularly round or sharp-toothed.

The scape is 3–10 cm. tall, but 20–50 cm. in fruit, and carries an umbel of 4–12 flowers. The pedicels are 1 cm. or less in flower, but up to 2 cm. in fruit and then stiff and erect. There may be several scapes to each plant.

The corolla is rose, pink, or even a good deep red, always with a yellow eye. There is no annulus. The corolla-tube is 1–2 cm. long, 1½–2 times the calyx. The limb is 1–2 cm. diameter with obcordate lobes deeply notched at the apex.

P. rosea is a plant of the north-western Himalaya: Garhwal, Kashmir and Afghanistan, where it grows in wet meadows, on the banks of and even in small streams, and on grassy alps beneath the melting snows. Its altitude is 2700–4000 m.

P. rosea is a completely herbaceous plant which, when the flower-buds push through the soil in early April, is a delightful patch of colour. But there is subsequently rather a lot of foliage. The plant is quite intolerant of dry conditions during its subsequent period of growth.

P. warshenewskiana Fedtsch. (Plate XVIII)

P. warshenewskiana is a dwarf efarinose or sparingly farinose plant with a short stout rhizome.

The leaves, including the petiole, are 1.5–7 cm. long and 15–18 mm. broad, oblong or obovate to oblanceolate, rounded or obtuse at the apex, at the base tapering into a short winged petiole which is rarely more or less equal to the blade. The margin is finely toothed.

The scape is very short (2–3 mm.) or 2–5 cm., and lengthens in fruit. It bears an umbel of 1–8 flowers. The pedicels are 3–10 mm., lengthening in fruit to 2.5 cm.

The annulate corolla is rose or pink with a yellow throat surrounded by a narrow, white, more or less 5-lobed area. The corolla-tube is about 1–1.5 cm. long, 2–3 times the calyx, and the limb is 7–12 mm. diameter with obcordate lobes deeply notched at the apex.

The distribution of this species appears to stretch from eastern Bokhara in Turkestan to the north-western Himalaya, where it grows in streamlets and the wet ground about springs at an elevation of 1700–2700 m.

It is a tiny plant, which lends itself to pan-culture, when the flowers may be better appreciated. But like most primulas, it grows better in the open ground, preferably in shade, and may be repotted again in autumn.

20 THE SECTION DENTICULATA

The Species of the Section

P. atrodentata	*P. erosa*	*P. laxiuscula*
*P. denticulata	*P. erythrocarpa*	*P. pseudodenticulata*

The inflorescence of species of this small section is usually a dense, many-flowered head, the flowers being sessile or nearly so, and more or less erect. The section lies between Capitatae and Farinosae. From Capitatae the section is distinguished by the erect flowers, which in

Capitatae (and Muscarioides) are pendant; and those species of the section Farinosae which have sessile flowers usually have pouched bracts, a feature which is lacking in Denticulatae.

The distribution of the section extends eastwards from Afghanistan, throughout the entire ranges of the Himalaya, and into southern Tibet, Assam, northern Burma, and Yunnan, Szechuan and Kweichow in China.

P. denticulata Smith (Illustration p. 23. Plate XIX)

This, the type species of the section, is a robust, finely hairy plant with a short, stout rhizome from which grow the thong-like roots. In mid-April, the still-contracted leaves part the earth to form a vortex through which grows the stout scape bearing a more or less globose head of flowers. This feature gives to the plant its garden name: 'the drumstick Primula'. The rhizome is clothed at the base with rather fleshy, broadly-ovate bud scales, which are 1.5–4 cm. long and 1–2 cm. broad.

The leaves form a compact rosette or tuft and, at flowering time, are 3–15 cm. long, including the petiole, and 1.5–4 cm. broad, but elongating at fruiting time to 20 cm. or more, in which condition the leaves are much more erect. The lamina is oblong to oblanceolate, rounded or obtuse at the apex, and at the base tapering into the broadly winged petiole, which may be very short or more or less equal to the lamina. The margin is toothed, and often recurved.

The scape is 5–30 cm. tall, but up to 45 cm. in fruit and there may be several scapes to a plant. The pedicels are 0–5 mm.

The corolla is of various shades of purple or reddish purple or occasionally white. There is no annulus. The corolla-tube is 0.5–2 cm. long, 1–2 times the calyx, and the limb is 1–2 cm. diameter with deeply divided obcordate lobes.

This, the only species of the section in cultivation, is widely grown and even used as a bedding plant. In flower it is a good garden plant, but later there is much foliage. It benefits from an annual dressing with bonemeal, and likes moisture and a little shade. It may be easily raised from seed and selected forms can be propagated by division. This should be done when spring growth is imminent (when the flowers will be lost) or after flowering. In either case the divisions may be replanted directly and well watered and shaded, particularly after flowering. To keep growth vigorous, a plant should be broken up at least every three years.

21 THE SECTION CAPITATAE

THE SPECIES OF THE SECTION

P. capitata   *P. glomerata*

KEY

Flowers pendant. Petals shallowly notched   *capitata*
Flowers erect. Petals deeply notched   *glomerata*

Like *P. denticulata* and its allies, the species of this small section occupy a position intermediate between Farinosae and Muscarioides.

Capitatae differ from Muscarioides in the glabrous leaves of harsher texture which are winter-persisent, but much reduced; in the globose and not spicate inflorescence; in the shortly-stalked but not sessile flowers; and in the greater regularity of structure of both calyx and corolla.

In Denticulata the flowers are erect, while in Capitatae the flowers are pendant, except in *P. glomerata* which has flowers more or less erect, but otherwise has the habit of Capitatae – and also the same flowering period, which is July-August for Capitatae, but April-May for Denticulata.

The distribution of Capitatae includes Nepal, Sikkim, Bhutan, southern Tibet and western Yunnan, but the section is absent from the western Himalaya, which is the main locus of Denticulata.

P. capitata Hook. (Illustration p. 24)

This species is a plant with a short or long, more or less stout rhizome which is without basal bud-scales.

The leaves, including the petiole, are 2–13 cm. long and 0.5–2 cm. broad, oblong or oblanceolate to oblong-spathulate, acute or rounded at the apex and tapering at the base into the winged petiole which may equal the lamina, but is usually much shorter. The lamina is sharply toothed, often recurved at the margin, and farinose to efarinose, with the upper surface less thickly coated with farina in farinose forms.

The scape is 10–45 cm. tall, several to a plant, each rising from the centre of a separate rosette of leaves growing along the rhizome. Each scape carries few to many deflexed flowers in a somewhat flattened globose head. The pedicels are very short, elongating in fruit to not exceeding 1.5 mm.

The corolla is blue-purple to deep purple with a yellow eye and both annulate and exannulate forms occur. The corolla-tube is 0.5–1.5 cm. long, 1–2 times the calyx, and the limb is 7–10 mm. diameter, with more or less spreading obcordate lobes which are shallowly notched at the apex. Both pin-eyed and thrum-eyed flowers occur.

The usual altitudinal range is 4000–5000 m.

P. capitata likes a moist leafy soil with a little, light shade. If grown in a harsh or dry soil, the rhizome tends to grow a little above the surface of the soil and the contracted, winter-green rosettes break off. This may suggest a natural method of vegetative propagation, but is not conducive to the growth of a good plant.

There is a particularly fine form of *P. capitata* in Ludlow and Sherriff 17507. The colour is a rich violet and it may be in flower from July to September, depending on the season.

P. glomerata Pax

This species is a plant with a short or long, rather stout rhizome which is without basal bud-scales, but is sometimes clothed with the sheathing-bases of old leaves.

The leaves, including the petiole, are 3–15 cm. long and 1–3 cm. broad, oblong or oblanceolate or obovate-spathulate, rounded at the apex, and tapered into the winged petiole which is usually red at the base. The petiole may be more or less equal to the lamina, but is usually much less. The margin is irregularly sharp-toothed and the lamina is often markedly white-farinose below when young.

The scape is 10–30 cm. tall and carries a compact, more or less globose head of flowers of which many are erect. The pedicels are 0–10 mm. long.

The corolla is blue, funnel-shaped, and usually exannulate, but may be annulate. The corolla-tube is 8–16 mm. long, 1–2 times the calyx. The limb is concave, 6–10 mm. diameter with obcordate lobes deeply notched at the apex. Both pin-eyed and thrum-eyed flowers occur.

P. glomerata has a fairly wide distribution in Nepal, Sikkim and in Tibet to the immediate north of Sikkim. The altitudinal range is given as 3300–5700 m. The species was brought into cultivation, subsequent to the importation of wild seed, in about 1930, but it is doubtful if it still persists.

THE SPECIES OF THE SECTION

P. aerinantha	*P. euchaites*	**P. muscarioides*
**P. apoclita*	*P. giraldiana*	*P. pinnatifida*
**P. bellidifolia*	*P. gracilenta*	**P. vialii*
P. cernua	**P. hyacinthina*	*P. violacea*
**P. concholoba*	*P. inopinata*	*P. watsonii*
P. deflexa	*P. mairei*	

KEY

1	Calyx crimson. Petals pointed	*vialii*
	Calyx not crimson. Petals notched or rounded	2
2	Leaves farinose beneath	*hyacinthina*
	Leaves efarinose	3
3	Corolla tube scarcely longer than calyx	*concholoba*
	Corolla tube at least 2 times calyx	4
4	Flowers 5–15 in a head, some held horizontally	*bellidifolia*
	Flowers 10–100 in a spike, all down-turned	5
5	Leaves less than 10 cm., hairy on both surfaces	*apoclita*
	Leaves mostly 10 cm. or more, almost glabrous below	*muscarioides*

The species of this section are closely allied to Capitatae and Soldanelloideae, between which sections they are placed.

In Capitatae, the flowers are arranged in a more or less globose head, but in Muscarioides the inflorescence takes the form of a spike. The shape and droop of flowers of both sections is more or less the same and in both sections the flowers develop from the base of the inflorescence upwards.

The flowers of Muscarioides are sessile, while those of Capitatae are shortly stalked. This is an important distinction. The downward curvature of the flowers of Muscarioides is produced by stronger development of those floral segments adjacent to the scape, particularly those of the calyx and corolla. This results in a slightly irregular flower. With Capitatae, downward curvature is produced by bending of the pedicel, and the flower is quite regular.

The shape of the corolla is usually sufficient to distinguish between species of the sections Muscarioides and Soldanelloideae. In Muscarioides, the corolla-tube is usually much longer and more conspicuous than the limb, which has short lobes more or less at right angles to the corolla-tube. In Soldanelloideae the corolla tube is short, and the enlarged limb widens into a broad bell with the lobes continuing the bell-shaped profile.

Species of the section Muscarioides are similar in general habit to Capitatae, but tend to be less robust and sometimes dwarf. The rhizome is only weakly developed, so that even in the wild they tend to be rather short-lived perennials. In cultivation, the more robust species tend to set seed freely. But the weaker species tend to die after flowering. A few of the more vigorous species therefore persist in cultivation, while the weaker members require the importation of fresh seed.

The distribution of the section includes the eastern Himalaya, Tibet and western China.

P. apoclita Balf.f.et Forrest

This species has leaves 3–5 cm. long and 1–1.5 cm. broad, oblong-spathulate to oblong or elliptic, rounded at the apex, and rapidly tapering into the winged petiole which is usually very short. The margin is regularly crenate-serrate to crenate-dentate and both surfaces of the lamina are lightly covered with hairs.

The scape is up to 16 cm. tall and carries a short spike or head of fragrant, reflexed flowers.

The corolla is deep purplish-blue and about 12 mm. long. The corolla-tube is 7 mm. long, 2 times the calyx and widening upwards into the limb. The lobes are 2–3 mm. long and 3–3.5 mm. broad, and faintly notched or shallowly round-toothed at the apex. Pin-eyed and thrum-eyed forms occur.

P. apoclita is a plant of moist, stony alpine pastures at an elevation of 4000–4700 m. Its distribution includes south-eastern Tibet, western Yunnan and south-western Szechuan.

The species was flowered in cultivation in 1924 and persisted for a number of years, but it is doubtful if it still remains in cultivation.

P. bellidifolia King ex Hook. (Plate XXI)

This species has leaves 4–15 cm. long and 1–2.5 cm. broad, oblanceolate to spathulate, rounded at the apex and narrowed at the base into the more or less hairy, winged petiole. The margin is shallowly sharp- or round-toothed. In shape and texture the leaves resemble those of the common daisy, whence the specific name.

[103]

The scape is 10–35 cm. high and carries a head of 7–15 reflexed flowers.

The corolla is mauve to pale blue-violet with a ring of farina at the throat. The corolla-tube is 5–10 mm. long, 1–2 times the calyx, and rather darker than the limb. The lobes are obovate and conspicuously notched at the apex. Both pin-eyed and thrum-eyed forms occur.

This species comes from the alpine slopes of Sikkim, south-eastern Tibet and Bhutan at an altitude of 4000–5300 m.

P. bellidifolia is rarely offered by nurserymen, but in 1975 viable seed was offered by the Scottish Rock Garden Club exchange. With scores of Primula and other delectable seedlings to be had, the dozen or so well-grown seedlings of this particular species were selected for attack by some low form of life, which removed the whole of the growing point. Slugs and woodlice are both suspect.

P. concholoba Stapf et Sealy

The leaves of plants of this species are 2–8 cm. long and 1–2 cm. broad, oblanceolate-oblong to oblong, rounded at the apex, long-attenuate at the base into the more or less winged petiole, and coarsely and irregularly toothed at the margin.

The scape is 7-20 cm. high and carries a compact, globose head of 10–20 flowers.

The corolla is bright violet and more or less powdered with snow-white farina outside. The corolla-tube is about 5 mm. long, scarcely longer than the calyx, cylindrical for the lower 3–4 mm., and then gradually widening into the more or less globose, bell-shaped limb. The corolla-lobes are concave, broadly elliptic to almost round, 1.5–3 mm. long, and entire or shallowly notched at the apex. This species is monomorphic, the style projecting slightly above the stamens.

This quite distinct species comes from the region of the Assam-Burma-Tibet frontier where it grows, at an altitude of about 4000 m., on steep grassy slopes and on cliff ledges amongst dwarf juniper and rhododendron.

P. hyacinthina W.W.Sm.

The leaves of plants of this species are 16–18 cm. long and 2–2.5 cm. broad, oblanceolate to oblong, obtuse or rounded at the apex, gradually tapering into the hairy, slightly winged petiole and irregularly toothed at the margin. The lower surface of the leaves is covered with white farina, *P. hyacinthina* being the only member of the section to possess this characteristic.

The scape is 20–45 cm. high and the numerous, deflexed flowers are arranged in a compact spike 2.5 cm. in length, or in a head 2–2.5 cm. diameter.

The violet corolla is about 17 mm. long, with a narrow tube about 10 mm. long, 1½–2 times the calyx. The lobes of the limb are obcordate, 4 mm. long and 5 mm. broad. Pin-eyed and thrum-eyed forms occur.

This species, which has a strong scent of hyacinth, grows at 4200–5300 m. on stony hillsides in south-eastern Tibet.

Seed is sometimes offered in India.

P. muscarioides Hemsl. (Illustration p. 25)

The leaves of plants of this species are obovate-spathulate or elliptic to oblong, 10–20 cm. long and 3–5 cm. broad, rounded at the apex and gradually tapering into the winged petiole. The margin is round or sharp toothed and the lamina is efarinose on both surfaces.

The scape is up to 40 cm. tall and carries the conspicuously reflexed, fragrant flowers in a compact head or a short spike.

The corolla is a deep purplish-blue and about 12 mm. long. The corolla-tube is 8 mm. long, about 1½ times the calyx. The lobes are 2 mm. long and 2 mm. broad, broadly elliptic, and truncate or shallowly notched at the apex. Pin-eyed and thrum-eyed forms occur.

This fine species grows in wet meadows. Its distribution extends from south-eastern Tibet to Szechuan and Yunnan.

P. vialii Delavay ex Franch. (Plate XX)

The leaves of *P. vialii* are broadly lanceolate to oblong, 20–30 cm. long and 4–7 cm. broad, rounded at the apex and tapering into the winged petiole. The margin is remotely and shallowly toothed or irregularly and sharply toothed. The lamina is efarinose but hairy on both surfaces.

The scape is 30–60 cm. tall and carries the many flowers in a dense spike 6–18 cm. long.

The corolla is bluish-violet and about 13 mm. long. The corolla-tube is about 7 mm. long, 1½ times the calyx. The lobes are 5 mm. long and 4 mm. broad, ovate to elliptic and acute at the apex. Pin-eyed and thrum-eyed forms occur.

The calyx of flowers in bud is bright crimson, and the lowest flowers of the spike open first. The immature spike therefore presents a striking and distinctive appearance: bluish-violet flowers in the lower part, capped by the more or less overlapping crimson calyces at the head of the spike.

P. vialii comes from north-western Yunnan and south-western Szechuan, where it grows at upwards of 3000 m. It is a late-flowering plant for a primula, July onwards. In cultivation it sets seed freely. A particularly robust form may be encountered under the synonym *P. littoniana*. This form was named for George Forrest's fellow collector, G. L. Litton, who died at Tengyueh in December 1905.

23 THE SECTION SOLDANELLOIDEAE

THE SPECIES OF THE SECTION

*P. buryana	**P. nutans*	*P. soldanelloides*
*P. cawdoriana	**P. reidii*	*P. spicata*
*P. chasmophila	*P. sandemaniana*	*P. umbratilis*
*P. eburnea	**P. sapphirina*	*P. uniflora*
P. fea	**P. sherriffiae*	*P. wattii*
P. flabellifera	*P. siamensis*	**P. wigramiana*
P. jigmediana	*P. siphonantha*	**P. wollastonii*
P. ludlowii		

KEY

1 Flowers predominantly white — 2
  Flowers blue to violet (with or without white eye
    or reverse) — 5
2 Flower funnel-shaped — 3
  Flower with globular limb — *reidii*
3 Corolla 2 cm. or less in length — 4
  Corolla about 3 cm. — *wigramiana*
4 Calyx hairy. Petal lobes 6–8 mm. with single deep
    notch — *buryana*
  Calyx hairless (farinose). Petal lobes 3–4 mm.
    long, irregularly notched — *eburnea*
5 Very small. Leaves to 1 cm. Flower 5–6 mm.
    across — *sapphirina*
  Larger. Leaves 2 cm. or much more. Flower 1 cm.
    or more across — 6
6 Flower funnel or bell shaped — 7
  Flower with globular limb — *reidii* var.
      williamsii

[106]

7 Leaves slightly bullate, narrowing abruptly to
    petiole, pinnately lobed   *chasmophila*
  Leaves not bullate, tapering to petiole, not lobed   8

8 Calyx hairy   *buryana*
  Calyx glabrous or farinose   9

9 Corolla greenish white on reverse. Petals narrow,
    cut into 2–3 lobules   *cawdoriana*
  Corolla blue to violet within and without. Petals
    entire or with single notch   10

10 Corolla tube 35 mm. or more   *sherriffiae*
  Corolla tube 10 mm. or less   11

11 Leaves more or less upright. Flowers funnel-
    shaped. Corolla tube 10 mm.   *nutans*
  Leaves in flat rosette. Flowers bell-shaped. Corolla
    tube 5 mm.   *wollastonii*

The Soldanelloideae are most closely akin to Muscarioides, which they resemble in habit and in the texture and appearance of the leaves. These are shortly and softly hairy, but never farinose.

The flowers of the Soldanelloideae are sessile and reflexed or nodding. They may be single, but are usually carried in a head or spike.

The chief distinction lies in the flower structure. In Soldanelloideae, the cylindrical base of the corolla-tube is usually short, and expands more or less rapidly into the much more pronounced bell-shaped or funnel-shaped limb. Further, the lobes are more or less in line with the tube, and not sharply bent back at right angles as with the limb of most primulas. In Muscarioides, the tube is the more prominent part of the corolla.

The Soldanelloid primulas are regarded as one of the more difficult sections in cultivation. They are also very beautiful plants. In the wild they are mostly high alpines, with a covering of snow for from four to six months of the year. This will keep the plants dry, and at an even temperature around freezing-point. In cultivation they like plenty of leaf-mould, a buoyant, moist environment and some shade. Many species will flower profusely and if this results in the production of viable seed, then the species will persist. But all too frequently seed is either infertile or aborted. Most species of the section are short-lived perennials of but little stamina and the plant, in the absence of freshly imported wild seed, is lost.

Mr. Drake, who gardens at Aviemore in the Scottish Highlands, makes a point with respect to the period of winter dormancy. In his district, there are periods of severe frost before the advent of snow, and he observed that the frost appeared to shatter the still green parts of species of this section. At the beginning of October, he therefore places a frame-light over his plants, which are quite dried out by the time the frosts arrive, and the plants then appear to suffer little damage. But even Mr. Drake has now only about a fifth of the species of this section that he formerly grew. And then there was that plant of *P. cawdoriana* seen growing at Alston in Cumbria. It was in a fairly open bed, got no winter protection, and its growth was like that of a small cabbage.

With three exceptions, the species of this section are confined to the ranges of the Himalaya. The exceptions are *Pp. nutans* and *spicata* from south-western China, and *P. siamensis*.

P. buryana Balf.f.

Leaves 3–5 cm. long, ovate or oblong-ovate, rounded at the apex and tapering into the winged petiole which is more or less equal to the blade. The margin is irregularly round or sharp-toothed, and both surfaces of the lamina are hairy.

The scape is 8–18 cm. tall and carries a head of 4–7 semi-pendant flowers.

The funnel-shaped corolla is 1.2–1.8 cm. long, white or rarely purple. The corolla-tube is 6-7 mm. long, more or less equalling the calyx, shortly and softly hairy, and gradually widening into the somewhat spreading limb. The lobes are 6–8 mm. long and 5–7 mm. broad, obovate and deeply notched at the apex. Both pin-eyed and thrum-eyed forms occur.

A species somewhat tenuously in cultivation.

P. cawdoriana Ward

The leaves are obovate, oblanceolate, or spathulate, 2–4 cm. long and 1–1.5 cm. broad, obtuse or rounded at the apex and tapering into the broad petiole which is 5–10 mm. long. The margin is coarsely and irregularly sharply-toothed, and both surfaces of the lamina are slightly rough to the touch. The leaves of plants of this species grow in a rather low, depressed rosette, in contrast to those of some species of this section which grow in a much more upright rosette, *P. nutans* for example.

The scape is 6–15 cm. tall and carries a head of 3–6 slightly depressed flowers.

The corolla is quite distinctive within the section. It is more or less narrowly conical, with the narrow lobes deeply notched and slightly spreading. The colour is white or greenish-white below and blue or violet to mauve above, including the limb. There is a large white eye. The corolla is about 3 cm. long, about 4 times the calyx, and the limb 2–3 cm. diameter. Long and short-styled forms occur.

P. cawdoriana is known only from south-eastern Tibet, where it was found by Ward in 1924, growing at 4000–4700 m., on steep, rocky slopes.

The species is in cultivation, and available, but it is rather scarce.

P. chasmophila Balf.f.ex Hutchinson (Plate XXII)

The slightly blistered leaves are oblong-lanceolate to oblong-ovate, 3–8 cm. long and 1–2 cm. broad, rounded or obtuse at the apex and abruptly narrowed into the winged petiole which equals or exceeds the blade. The margin is more or less pinnately lobed and both surfaces of the lamina are shortly and softly hairy.

The scape is 6–9 cm. tall, and carries 1–5 slightly pendant flowers.

The funnel-shaped corolla is about 2 cm. in both length and diameter, 3–4 times the calyx. Colour is deep violet. The lobes are obovate to orbicular and rather widely notched at the apex. Long and short-styled forms occur.

This species comes from Bhutan, where it grows in peaty rock-crevices at 5000–5500 m. It is a species which has died out of cultivation, but it could be re-introduced.

P. eburnea Balf.f.et Cooper (Plate XXIV)

The leaves are elliptic to oblong or ovate, 3–9 cm. long and 1.5–3.5 cm. broad, rounded at the apex and gradually tapering into the winged petiole, which is more or less equal to the blade. The margin is sharp-toothed or lobed.

The scape is 10–20 cm. high, and bears a compact head of 6–12 flowers.

The funnel-shaped corolla is 1.5 cm. long, about 2 times the calyx, and 1 cm. diameter. Colour is ivory white, with no eye. The lobes are 3.5–4 mm. long, broadly rounded and irregularly notched. The bases of the lobes are somewhat distant, but the apices are more or less touching. Short-styled and long-styled forms occur.

P. eburnea grows at 4300–4800 m. on peaty turf, glacial debris and at the foot of cliffs in Bhutan. Like *P. chasmophila*, this species appears to be in limbo, but could be reintroduced, and Sherriff (in 9,

p. 35) describes the species as 'the gem of the whole of the eastern Himalaya'.

P. nutans Delavay ex Franch. (Illustration p. 26)

The leaves of plants of this species are narrowly elliptic to broadly oblanceolate, 5–20 cm. long and 2–5 cm. broad, rounded or obtuse at the apex and gradually tapered into the winged petiole which equals one half to one third of the blade. The lamina is efarinose and the margin wavy, with sharp teeth.

The scape is 10–50 cm. high, with 5–15 downward pointing flowers usually carried in a short spike 2–5 cm. long but the inflorescence of fewer-flowered forms may be a dense head.

The lavender to violet funnel-shaped corolla is 2–2.5 cm. long. The lower, constricted part of the corolla-tube is 1 cm. long, about 2 times the calyx. The limb is 2–2.5 cm. diameter with ovate or suborbicular lobes which are sometimes entire, but usually shallowly notched at the apex, often with a minute tooth at the bottom of the notch. Long-styled and short-styled forms occur.

This species was discovered by Delavay, growing in open pine forests and rocky pastures at 3500 m. It has a wide but apparently discontinuous distribution in the ranges of western Yunnan. It also occurs within the adjacent marches of south-western Szechuan, and in eastern Yunnan.

P. nutans is much the most vigorous species of the section. It is an open ground plant and will set seed in favourable seasons. Smith and Fletcher (7) say that this is a monocarpic species, but this is not invariably the case. Viable seed of most primulas does not readily set in my climate and I therefore usually take the scape off the plant of most species soon after flowering. Plants of *P. nutans* treated in this way certainly persist for a few seasons, and I have even divided a robust plant of *P. nutans* into three pieces and grown these on in open ground.

P. reidii Duthie (Plate XXV)

The leaves are oblong to oblong-lanceolate, 5–20 cm. long and 2–3 cm. broad, obtuse or rounded at the apex, and gradually tapering into the slightly winged petiole which nearly equals the blade. The margin is shallowly lobed and toothed to coarsely round-toothed. Both surfaces of the lamina are efarinose, with some long hairs.

The scape is 6–15 cm. tall and carries a compact head of 3–10 pendant flowers.

The corolla is ivory white and 1.8–2.2 cm. in length and diameter. The corolla-tube is 9–10 mm. long, 1–1½ times the calyx, and expands abruptly into the almost globular limb. The lobes are broadly oblong or ovate, and shallowly notched at the apex, with or without a tooth in the notch. Long-styled and short-styled forms occur.

This is a beautiful little plant. It grows at 4000–4300 m. in Garhwal and Kumaon in the north-western Himalaya. Its habitat is described as wet rocks.

P. reidii var. *williamsii*. This is a slightly more robust form, introduced in 1952–54 from western and central Nepal by L. H. J. Williams of the British Museum (Natural History). It is also possibly an even more beautiful plant, with wonderfully fragrant, gorgeous bells which are pale blue to white or white in the lower part.

These plants will germinate like mustard and cress from viable seed, provided there is a good proportion of leafmould in the compost. They are entirely herbaceous and may be grown in open ground provided some shade is given. They must also be kept moist during the growing season. Dryness rapidly kills them. Being small plants, they may also be grown in pots. But they are not long-lived plants and a succession should be maintained from seed.

P. sapphirina Hook.f.et Thoms.

The leaves are 5–10 mm. long and 2–5 mm. broad, oblanceolate to obovate and tapering into a short, broad petiole. The margin is coarsely toothed and there are short white hairs on the upper surface of the lamina.

The thread-like scape is up to 5 cm. tall and carries 1–4 semi-pendant flowers.

The funnel-shaped corolla is 4–5 mm. long, 1½–2 times the calyx. Colour is violet-purple to blue. The corolla-tube is very short, and the limb is 5–6 mm. diameter. The ovate lobes are deeply divided. Long-styled and short-styled forms occur.

This very dwarf species grows at 4000–5000 m. in the Himalaya of Sikkim, Bhutan and the adjoining parts of south-eastern Tibet. Its habitat is damp, rocky ground and even mossy boulders, with no soil contact and moisture provided by the frequently-swirling mists.

Seed is occasionally offered in India.

P. sherriffiae W.W.Sm.

The efarinose leaves are 6–17 cm. long and 2.5–5 cm. broad, usually oblanceolate but sometimes elliptic, or again more or less ovate, rounded at the apex and narrowed into the winged petiole which is 1–2.5 cm. long. The margin is entire or remotely toothed and fringed with white hairs. The leaves grow in the form of a flat rosette.

The scape is 4–10 cm. tall and carries a compact head of 2–7 flowers, which are almost horizontal.

The corolla is pale violet with a white edge, and flecks of farina outside and at the throat inside. It is about 5 cm. long and the expanded limb is 2–3 cm. diameter. It is shaped like a long funnel. The corolla-tube is 3.5–4 cm. long, 3½–4 times the calyx, and 1.5–2 mm. diameter, slightly curved and suddenly expanded near the apex. The lobes are more or less orbicular and about 1 cm. diameter. The flowers of this species are monomorphic, with the style shortly exceeding the tips of the anthers.

P. sherriffiae was found growing on cliffs in south-eastern Bhutan at about 1700 m. It is not regarded as hardy out of doors (note the low elevation) but it can be grown in pots in the cool alpine house.

The species appears to be in cultivation, but rare. It is not offered by nurserymen as yet, but there are signs that *P. sherriffiae* is making slow progress and should be looked for.

P. wigramiana W. W. Sm.

The leaves are oblanceolate to oblong or ovate, 6–10 cm. long and 2–3.5 cm. broad, rounded at the apex, and narrowed into the winged petiole which is 2–4 cm. long. There are many soft hairs on both surfaces of the lamina and the margin is remotely toothed. In habit, the plant forms a spreading rosette.

The scape is 20–25 cm. tall and carries a head of 6–7 pure white pendant flowers.

The corolla is widely funnel-shaped and about 3 cm. long. The constricted part of the tube is 10–12 mm. long, slightly longer than the calyx, and expands into the limb which is 2.5 cm. diameter. The lobes are round to ovate, 1 cm. long and usually distinctly toothed, but sometimes more or less entire. There are long-styled and short-styled forms.

P. wigramiana was first found in 1931 by Professor K. N. Sharma, growing at 5700 m. in the mountains of Nepal. Smith and Fletcher (7) describe *P. wigramiana* as a 'very fine species', but it is doubtful if it still remains in cultivation. This is rather surprising, as the species appears to be one of the more robust members of the section. Is it too much to hope for its reintroduction?

P. wollastonii Balf.f.

The leaves are oblanceolate to obovate, 2.5–5 cm. long and 1.25–2.5 cm. broad, rounded at the apex and narrowed into the short, winged petiole. The margin is sharp or round-toothed, with dense, regularly arranged hairs projecting from the margin. There are white hairs on both surfaces of the lamina, which is efarinose or with copious farina below. The habit of the leaves is that of a flat rosette.

The scape is 12–20 cm. high and carries a head of 2–6 pendant flowers.

The corolla is bell-shaped, 1.8–2.5 cm. long, dark purple to bright blue, and very farinose to more or less efarinose both within and without. The lower, cylindrical part of the corolla is 5 mm. long, about equal to the calyx, and rapidly expands into the limb. The lobes are 5 mm. long and broadly ovate and entire. There are long-styled and short-styled forms.

This species was first collected during the 1921 Everest expedition by Dr. A. F. R. Wollaston. It was found at 4600–5000 m. in southern Tibet. It was subsequently found in 1930 by Major Lal Dhwoj in eastern Nepal. These latter plants were much more farinose.

In cultivation, *P. wollastonii* rarely sets seed and is probably not now in cultivation. But it might be reintroduced, and Smith and Fletcher (7) say that it can be successfully propagated by means of root cuttings.

24 THE SECTION ROTUNDIFOLIA

THE SPECIES OF THE SECTION

P. baileyana	*P. consocia*	*P. ramzanae*
P. barnardoana	**P. gambeliana*	**P. rotundifolia*
P. caveana	*P. littledalei*	*P. tzetsouensis*

KEY

Inflorescence efarinose. Petals deeply notched	*gambeliana*
Inflorescence farinose. Petals entire or shallowly round-toothed	*rotundifolia*

The species of this small section approach the large Nivalid aggregate in the structure of both calyx and corolla, and in the capsule and seeds. But the species of the section differ conspicuously from the Nivalids in the leaves which, in the latter section, are typically strap-shaped. In the section Rotundifolia the leaves have a more or less rounded blade with cordate base and a long, slender, distinct petiole.

P. gambeliana Watt

This is a perennial species with a very short rhizome and long fibrous roots. The rhizome is clothed at the base with a few, overlapping ovate white- or yellow-farinose bud-scales.

The leaves, including the petiole, are 2–30 cm. long. The lamina is 1–10 cm. long and 0.5–6 cm. broad, ovate to orbicular; rounded, obtuse or acute at the apex, usually deeply cordate at the base, but occasionally truncate. The lamina is efarinose and the margin sharp or more or less bluntly toothed. The petiole is up to four times the length of the lamina, and quite distinct.

The scape is 3–25 cm. long, shorter than the leaves to twice their length, and carries an umbel, or sometimes two superposed umbels, each with 1–8 flowers. The erect pedicels are 2–10 mm. long but up to 2.5 cm. in fruit. The inflorescence is efarinose.

The annulate corolla is purple-pink to violet-purple with a yellow throat. The corolla-tube is 1–2.5 cm. long, 2–3 times the calyx. The limb is 1.5–2.5 cm. diameter with broadly obcordate lobes deeply notched at the apex. Long-styled and short-styled forms occur.

P. gambeliana grows at an elevation of about 4500 m. It often grows among moss, or in the crevices of steep, wet cliffs. Its distribution is south-eastern Tibet, Bhutan, Sikkim and Nepal.

Seed is occasionally offered in India.

P. rotundifolia Wall. (Illustration p. 28)

P. rotundifolia is a perennial with a short, stout rootstock which is usually clothed at the base by overlapping, ovate to ovate-oblong, creamy-yellow farinose bud-scales and sometimes by the withered remains of old foliage.

The leaves, including the petiole, are 5–20 cm. long. The lamina is 2–12 cm. long and broad, cordate-reniform to cordate-orbicular. The margin is irregularly and coarsely round or sharp-toothed. The lamina is somewhat fleshy. The upper surface is without hairs and even shiny. The lower surface has copious creamy-yellow farina, or is often efarinose. The stout petiole is 1–3 times the lamina.

The stout scape is 10–30 cm. tall and carries a single umbel, or two superposed umbels, each with 2–16 flowers. The pedicels are 1–2 cm. at flowering time, but up to 3 cm. in fruit. The inflorescence is farinose.

The annulate corolla is pale purplish-pink with a golden yellow eye, and 1.5–2 cm. long. The corolla-tube is 1–1.5 cm. long, about 3 times the calyx. The limb is 1.5–2 cm. diameter, with obovate or orbicular lobes which are entire or more or less shallowly round-toothed at the margin. Long-styled and short-styled forms occur.

The distribution of this species is confined to the frontier areas of Nepal and Sikkim, where it grows at an elevation of 4000–5000 m. This is a difficult species to grow, but it is in cultivation, if somewhat tenuously, and seed is occasionally offered in India. Like *P. gambeliana*, this species can vary considerably in vigour. If grown in a pot, then the leaves require frequent spraying with water.

25 THE SECTION NIVALES

THE SPECIES OF THE SECTION

SUBSECTION EU-NIVALES

P. amabilis	*P. helvenacea*	*P. orbicularis*
P. bayernii	*P. kiuchiangensis*	*P. poluninii*
P. boreio-calliantha	*P. limbata*	*P. purdomii*
P. brevicula	*P. longipes*	*P. rigida*
P. cerina	**P. macrophylla*	*P. russeola*
*P. chionantha	**P. melanops*	**P. sinoplantaginea*
P. crocifolia	*P. minor*	**P. sinopurpurea*
P. diantha	**P. nivalis*	**P. stuartii*
*P. duthieana	**P. obliqua*	*P. tschuktschorum*
P. elongata	**P. obtusifolia*	*P. woodwardii*
P. farreriana	*P. optata*	*P. youngeriana*

SUBSECTION CALLIANTHA

P. bryophila	**P. calliantha*	*P. mishmiensis*

SUBSECTION MAXIMOWICZII

P. advena	**P. maximowiczii*	*P. tangutica*
P. aemula	**P. szechuanica*	*P. yüana*
P. handeliana		

SUBSECTION AGLENIANA

*P. agleniana	*P. elizabethae*	*P. falcifolia*

KEY

1 Efarinose. Petals narrow to strap-shaped, reflexed 2
  Usually farinose. Petals more or less rounded 3

2	Flowers generally dark purple or red	*maximowiczii*
	Flowers pale yellow	*szechuanica*
3	Corolla bell-shaped. Flowers exannulate	4
	Corolla with disc-shaped limb. Flowers annulate	5
4	Leaves 10–20 cm., regularly toothed. Corolla oblique with some reflexed petals	*obliqua*
	Leaves 20–30 cm., deeply indented. Corolla regular	*agleniana*
5	Corolla tube inflated	*calliantha*
	Corolla tube not inflated	6
6	Flowers purple to mauve, or white	7
	Flowers yellow	15
7	Petals deeply notched	*farreriana*
	Petals not or shallowly notched	8
8	Flowers white	*chionantha*
	Flowers purple to mauve†	9
9	Young leaves farinose	10
	Young leaves generally efarinose	*nivalis*
10	Farina white	11
	Farina yellow	13
11	Leaves rounded at apex. Flowers with a white or yellow 'eye'	*obtusifolia*
	Leaves more or less pointed. Flowers with dark 'eye'	12
12	Leaves usually less than 12 cm. Flowers with dark purple 'eye'	*macrophylla*
	Leaves usually more than 13 cm. Flowers with black 'eye'	*melanops*
13	Flowers large (up to 3.5 cm.)	*sinopurpurea*
	Flowers smaller (about 2 cm.)	14
14	Flowers with grey 'eye'	*sinoplantaginea*
	Flowers with dark purple 'eye'	*macrophylla*
15	Young leaves yellow-farinose below	*stuartii*
	Young leaves efarinose below or with narrow marginal band of white farina	*duthieana*

† P. *macrophylla* may 'sport' white.

In its distribution, the section Nivales is almost wholly Asiatic, reaching its maximum development in the mountains of western

China and in the Himalaya. The various species are plants of the high alpine meadows, particularly wet meadows. They also grow among rocks by the stream-side. But they are not bog plants.

The Nivalid primulas usually over-winter as a resting bud, protected by fleshy scales to the base of which are attached the persistent, brown-withered remains of the leaves of the preceding season. The more robust species can be propagated by careful separation of these buds and immediate replanting in open ground. This should be done in spring, when growth is just commencing.

As a section, the Nivalid primulas do not like pots. The deep-thrusting thong-like roots are too vigorous. One is, of course, always reluctant to put a single rare plant out of doors, and a first-year seedling may be overwintered in a pot to give better control against rotting of the crown.

The leaves are all radical and more or less strap-shaped, with the lamina gradually tapering into the petiole. The petiole is usually short or absent and it is frequently impossible to differentiate between petiole and lamina. The margin of the lamina is finely toothed or more or less entire.

The stout scape carries an umbel of many flowers which may be purple or yellow, rarely white. In contrast to the section Candelabra, the Nivalid primulas tend to form only a single umbel but with some species, *P. chionantha* for instance, there may be two or more umbels superposed.

The funnel-shaped corolla takes the form of a cylindrical tube, at the mouth of which the five petal-lobes overlap to form the flare or limb of the funnel.

In the wild, many of these primulas are fine, robust plants, but they do not take kindly to cultivation. Successful flowering of a plant depends on the inherent resistance of the species to British winters (and British summers in the south). It also depends on the skill with which winter protection is given to the resting bud and on the simulation of natural growing conditions. Again one must stress the fact that it is easier to provide these conditions in the cooler north of England and in Scotland than is possible in the south of England.

The section Nivales is divided into four subsections. The first subsection, Eu-Nivales, is based on the type species *P. nivalis*. The flowers of species of this subsection are generally purple (rarely yellow or white) and most (but not all) species are farinose, with silvery-white farina which diminishes or disappears after flowering. This subsection is much the largest and most widespread, and may be attributed to three geographical areas. The type species *P. nivalis* and its allies are found in central Asia, and extend to the north and east into Siberia and Alaska. In the Himalaya the chief representative of this subsection is *P. macrophylla* which, with some degree of geographical variation, occurs throughout the entire length of the range. In western and southern China, where there has been a greater degree of specific evolution, *P. sinopurpurea* may be taken as the median type.

The second subsection is based on *P. calliantha*. Species of this section are usually farinose, the farina being of a distinctive shade of greenish yellow. The flowers are purple or sometimes yellow and the corolla lobes are shallowly notched at the apex. This subsection is confined to south-western China and the adjacent parts of south-eastern Tibet. It also extends into north-western Burma and into the extreme eastern part of the Himalaya.

The third subsection is based on *P. maximowiczii*. The flowers are yellow, reddish or dark purple. Species of the subsection are usually efarinose. The corolla lobes are narrow, often linear, and more or less reflexed. Otherwise, the species of this subsection bear a general semblance to the Eu-Nivales. The subsection is restricted to northern and western China and to the adjacent parts of Tibet.

The fourth subsection is based on *P. agleniana*. Here the corolla is bell-shaped rather than funnel-shaped. The capsule also differs considerably from that of the three previous subsections, but growth and form is otherwise in general conformity with that of species of the other subsections of Nivales. This subsection is restricted to the area where converge the frontiers of China, Tibet and Burma.

The Nivalid primulas are difficult to grow; most of them indeed are wellnigh impossible. When seed is available it usually germinates well. But growth often does not progress beyond the seedling stage. With some species, a very few flowering plants have been raised – often, in the first instance, by the Royal Botanic Garden, Edinburgh – perhaps to linger another year, and then die. Some species have even gained an Award of Merit and then been lost.

For the successful establishment of most species of Nivalid primula, it would appear that viable seed must be available – and this is not produced by a single plant. The Nivalids are not self-sterile, but they should be grown in groups. Wild seed is more than usually difficult to obtain at the present time.

The story of success and despair with the Nivalids is written in the pages of the *Bulletin* – particularly the pages of the earlier volumes. At the present time it is probably true to say that a few species of the subsection Eu-Nivales alone are firmly established. But if there remain any pockets of Nivalid persistence – and surely there are some – perhaps their fortunate owners could be persuaded to contribute to the seed exchanges.

SUBSECTION EU-NIVALES

P. chionantha Balf.f. et Forrest (Illustration p. 29)

When growth first appears in spring, the leaves are strongly white-farinose. The upper part of the leaf is erect and triangular, and a few plants of *P. chionantha* then make quite a striking patch of vegetation. As the season progresses, the farina disappears and the now whitish-green leaves take on the long, thin, strap-shaped form typical of the Nivalid primulas, with the blade tapering gradually into the petiole. The apex is obtuse or rounded and the margin finely sharp-toothed or more or less entire. The leaves, including the petiole, are then about 15–25 cm. long and 2–5 cm. broad.

From each crown of a plant rises a stout to very stout scape 35–70 cm. high, which carries 1–4 umbels of numerous flowers. The pedicels are 2–3 cm. long and more or less nodding at first, but erect in fruit and then twice as long.

The annulate corolla is 1.5–2 cm. long and the limb 2.5–3 cm. diameter, with more or less obovate lobes which are entire. The colour is milky white with a yellow eye and there is a delightful vanilla fragrance. Both pin-eyed and thrum-eyed forms occur.

P. chionantha is a plant of open alpine meadows. It appears to be confined to north-western Yunnan, where it grows at an altitude of about 4000 m.

The affinity of *P. chionantha* lies with *P. sinopurpurea*, of which indeed it may well be no more than a white-flowered variety. Nor can any demarcation be made geographically, for *P. sinopurpurea* is a wide-ranging plant which is also found in north-western Yunnan. But *P. chionantha* is the more robust plant. It is more drought-resistant and altogether a better doer.

The first successful flowering of *P. chionantha* was at Kew in 1917, from seed obtained by Forrest in 1914. Since that time, the species has remained in cultivation. It sets fertile seed, from which it may be raised. The winter crowns may also be successfully divided in spring, when growth first commences. The divisions may be re-planted directly in the open ground, but should be kept moist.

P. duthieana Balf.f. et W. W. Sm.

This is a west Himalayan species and appears to have been found for the first time in Hazara in 1897 by one of Duthie's collectors. It was subsequently found in 1939 by R. L. Holdsworth, a member of this society, who describes the species (*Bulletin* Vol. 8, p. 25) as . . . 'a pale yellow primula, very luxuriant and robust in growth with masses of typical (strap-shaped) Nivalid leaves and flowers rising in loose clusters of up to thirty, from a scape that sometimes exceeds a foot in height. It grows at from 4000 to 5000 m., not only on cliffs over and through which trickle streams of melting snow, but also in softer places.'

Sadly, never in cultivation.

P. macrophylla Don

This species was first collected (as herbarium material) by Wallich in 1820. It is closely allied to the type species, *P. nivalis*, and has from time to time been regarded as a variety of *P. nivalis*. There are, however, various features which distinguish between the two species, and while none of these features is individually constant, collectively they establish grounds for separation. Further, observation of the two species when grown together in cultivation supports the need for separation. And in the wild they do not grow together: *P. nivalis* is a central Asian plant while *P. macrophylla* is Himalayan.

The west Himalayan plant is usually a stout species, although there are high-alpine forms of dwarfer stature. The species is a typical Nivalid, with firm strap-shaped, radical leaves which, including the petiole, are 15–25 cm. long and 1.5–3 cm. broad. The leaf blade tapers gradually into the broadly-winged petiole and it is usually not possible to differentiate between leaf-stalk and blade. The apex is acute, obtuse, or rounded, and the margin is finely round-toothed. The leaves are farinose to efarinose, but are usually white-farinose, at least on the underside of the leaves.

The scape is 12–25 cm. high and carries a single umbel of 5–25 flowers. The pedicels are 1–3 cm. long, but up to 8 cm. in fruit.

The annulate corolla is 1–3 cm. long and the limb about 2 cm. diameter, with more or less obovate lobes which are entire. Colour is purple, violet or lilac, with a dark purple eye which is rarely tinged yellow.

P. macrophylla is a plant of the mountain meadows and is found throughout the higher ranges of the Himalaya. It is however much more common and fully developed in the Western Himalaya, where the rainfall pattern diverges from the Monsoon and precipitation is much reduced. Holdsworth (1) describes three forms (as *P. nivalis* var. *macrophylla*) from the Kashmir-North West Frontier Province Border 'growing in rather exacting conditions on high mountains, where snow water is continually percolating during the period of growth and flower'. The species has a considerable altitudinal range, 4000–5300 m., and dwarf alpine forms occur as with many other primulas.

The typical form of *P. macrophylla* is more or less confined to the western Himalaya. In Nepal, Sikkim, Bhutan and the eastern Himalaya this is mostly replaced by the var. *macrocarpa*, a dwarfer form, and this plant does not extend westwards. Thus, in the Himalaya, from west to east there is a continuously changing form which culminates in western China in different but related types which have specific status.

Plants of this species can be difficult to establish. The form H78 (Huggins) is, reputedly, more amenable to cultivation. This plant has loose heads of lilac-purple flowers with a maroon-purple eye – an exceptionally lovely plant, with a delightful scent.

P. melanops W.W.Sm. et Ward (Plate XXVI)

The leaves are similar to those of *P. chionantha*, but narrower. The scapes are 20–35 cm. high, 1–3 rising from each plant and each scape carries one or two umbels of 5–12 flowers. The pedicels are 1–4 cm. long and somewhat nodding at first, but stiffly erect and elongated in fruit.

The corolla is about 2 cm. long and the limb 2 cm. diameter, with ovate lobes obtuse at the apex, and margin entire. Colour is purple to deep violet, with a black eye which gives the plant its specific name. Both pin-eyed and thrum-eyed forms occur.

The affinities of *P. melanops* are with *P. sinopurpurea*, and it also bears a strong resemblance to the Himalayan *P. macrophylla*. The latter, however, is geographically distant.

P. melanops tends to be rather short-lived but the species sets viable seed, from which plants may be raised. These should be potted and plunged in the alpine house or frame during the first winter, for planting in open ground the following spring, or planted directly in the open ground in a shady position.

P. nivalis Pallas

This is the type species of the section. It was found by Pallas as long ago as 1772 in Central Asia on the Altai Mountains. From here, the species extends northwards into southern Siberia and southwards across the Tien Shan in central eastern Sinkiang to the Alatau in south-eastern Russia. With this distribution, the plant endures a continental climate – much reduced precipitation and extremes of temperature.

The efarinose leaves are all radical, typically strap-shaped and about 15–25 cm. long and 1–4 cm. broad, with long, winged petioles.

The scape is 10–40 cm. high, elongating in fruit. The inflorescence is a single umbel of 3–25 flowers. The pedicels are 7–15 mm. long, elongating to twice that length in fruit.

The annulate corolla is violet to purple. The corolla-tube is 7.5–15 mm. long and the limb is 1.5–2.5 cm. diameter, with oblong, entire lobes. Both pin-eyed and thrum-eyed forms occur.

This species is recorded as having been cultivated in Europe since 1790, but only occasionally, for it is no more amenable to cultivation than other members of the section.

Seed is rarely available, and then doubtfully true.

P. obliqua W. W. Sm. (Plate II)

This species is fairly frequent in certain parts of Sikkim, where it grows on damp hillsides and in rhododendron thickets, at elevations upwards of 4000 m. In its distribution it extends to the west into Nepal, where also is found *P. stuartii*, a species with which *P. obliqua* was long confused. To the east, *P. obliqua* extends into Bhutan and the adjacent parts of Tibet, where it approaches the area occupied by *P. agleniana* and here *P. obliqua* was confused with *P. agleniana*. But the three species, *Pp. stuartii, obliqua* and *agleniana* are really quite distinct in form and growth.

The leaves are typically Nivalid with yellowish farina below and lamina tapering into the membranously winged petiole. The apex is usually obtuse and the margin regularly saw- or round-toothed. Length, including the petiole, is 10–25 cm. and width 2–5 cm.

The stout scape is 30–45 cm. tall and carries a single umbel of about 5 flowers. The pedicels are 1–3 cm. long.

The corolla varies from pale yellow to white, which may be rose-flushed. The corolla-tube is 1.5–3 cm. long and the limb 2.5–3 cm.

diameter. There is no annulus. Both pin-eyed and thrum-eyed forms occur.

One or more of the petal lobes is reflexed. There appears to be some difference of opinion as to which and how many petals have this feature. This is probably due to the examination of plants at different stages of flower development. This irregularity or obliquity of flower gives the plant its specific name.

P. obliqua has been in cultivation from time to time since 1918. But it is not a long-lived plant and fresh wild seed must be obtained for its continuation. In 1975 seed was brought back from Nepal by Beer.

P. obtusifolia Royle

The leaves are more or less oblong, rounded at the apex and tapering into the broadly winged petiole below. The margin is sharp- or round-toothed. The petiole is widely sheathing at the base and there of a reddish colouration. Length, including the petiole, is 10–30 cm. and width 3–10 cm. The leaves are of a thin texture and covered below with white farina.

P. obtusifolia is a plant of very variable stature. The scape is 10–50 cm. tall and carries from one to three umbels of 2–12 flowers. The pedicels are 1–4 cm. long.

The annulate corolla is purple or bluish-purple with a white or yellow eye. The corolla-tube is 1–2.5 cm. long and the limb to 2.5 cm. diameter. The petal lobes are obovate, with margin entire. Both pin-eyed and thrum-eyed forms occur.

The species is somewhat tenuously in cultivation. It was, for instance, offered by one nurseryman in 1968 and not again until 1975.

This species was first collected by Royle in 1831 at Kunawar in the north-western Himalaya. It is a plant confined to the western Himalaya, where it grows at an elevation of 4000–5300 m. on boulder screes in what are described as comparatively dry situations. This, however, may be misleading, as screes often carry water percolating from a higher altitude.

P. sinoplantaginea Balf.f.

At flowering time, the narrowly lanceolate leaf-blade is 6–9 cm. long and about 1 cm. broad. The blade is lightly covered below with pale yellow farina, finely round-toothed and always recurved at the margin, acute to obtuse at the apex and tapering at the base into the broadly-winged petiole.

The scape is 10–20 cm. tall, but elongates in fruit. The inflorescence is a single umbel of 5–12 fragrant flowers. The pedicels are 1–2 cm. long.

The corolla is a bluish to deep purple with a grey eye and markedly annulate. The corolla tube is 1–1.5 cm. long and the limb up to 2 cm. diameter, with oblong-elliptic entire lobes. Both pin-eyed and thrum-eyed forms occur.

The distribution of *P. sinoplantaginea* includes north-western Yunnan, western Szechuan and the adjacent areas of Tibet. It is a true mountain plant, growing at an altitude of 4600–5000 m.

This is not a long-lived species in cultivation. It has persisted somewhat intermittently since seed was first sent back by Forrest in 1914. It is related to *P. sinopurpurea*, which it resembles in general habit and with which it tends to hybridize in cultivation. *P. sinoplantaginea*, however, is much the smaller plant. The true species may now be lost, due to hybridization.

P. sinopurpurea Balf.f. (Plate XXVII)

P. sinopurpurea was long confused with the Himalayan *P. macrophylla* and with *P. nivalis* (as var. *purpurea*). It is however, geographically remote from *P. nivalis*, although forms bridging between *P. macrophylla* and *P. sinopurpurea* may yet be found. The affinity with *P. chionantha* was described under that species.

P. sinopurpurea is widely distributed and common in the ranges of western Yunnan at an elevation of 3000–4000 m. It also extends northwards into south-western Szechuan and to the west into the Tibetan marches north of eastern Bhutan.

P. sinopurpurea was first raised in 1911 from seed sent back by Forrest. It sets viable seed and has since remained in cultivation, although at times it has been scarce. It is less amenable to cultivation than is *P. chionantha,* requiring greater attention to watering and to protection during winter.

There is a noticeably long neck, to which are attached the persistent, brown-withered remains of old leaves. The leaf-blade is 5–35 cm. long and 1.5–5 cm. broad, oblong-lanceolate to very elongate, acute or obtuse at the apex and at the base gradually tapering into the broadly and membranously winged petiole. The margin is finely sharp-toothed and the lower surface is heavily coated with farina.

The scape may be only 10 cm. high, but robust forms in fruit may attain 75 cm., so that the species varies considerably in vigour. The scape carries 1–3 umbels with few to many flowers in each. The pedicels are 1–2.5 cm. long, somewhat reflexed in flower, but stiff and erect in fruit and then lengthening up to 8 cm.

The annulate corolla is purplish-violet with a white or grey eye. The cylindrical corolla-tube is up to 1.5 cm. long and the limb 2.5–3.5 cm. diameter, with more or less obovate, entire lobes. Both pin-eyed and thrum-eyed forms occur.

P. stuartii Wall.

This is a vigorous species. The oblong-lanceolate leaves, including the petiole, are 15–35 cm. long and 2–4 cm. broad, gradually tapering into the broadly winged petiole and covered below with a fine, yellowish farina. The margin is sharply-toothed.

The scape is 20–30 cm. high, but may elongate to 60 cm. in fruit. It carries one, and sometimes two, superposed umbels of numerous flowers. The pedicels are 2–3 cm. long.

The annulate corolla is golden yellow. The corolla-tube is 1.5–1.8 cm. long and the limb about 2.5 cm. diameter. The rounded lobes may be shallowly toothed at the margin or often entire, but not notched at the apex.

The area of distribution is Nepal and the N.W. Himalaya to the immediate west of Nepal, where it grows at an altitude of 4000–5000 m. Over this area it is a fairly uniform plant and could only be confused with *P. duthieana*. Since its first introduction to cultivation in 1847 (in Edinburgh) it has been intermittently available, but always rare. In 1974 seed was available in India.

SUBSECTION CALLIANTHA

P. calliantha Franch.

This species was discovered in 1883 by the Abbé Delavay, growing near Tali at an elevation of about 4000 m. It has subsequently been found on a number of occasions, and is always described as a very beautiful species. It is a plant that is locally abundant, but not widespread. The distribution is western Yunnan to south-eastern Tibet and north-western Burma.

The scape is 15–30 cm. tall and carries a single umbel of 3–10 large flowers. The markedly annulate corolla is deep purplish-violet to soft purplish-rose with a yellow-farinose eye. The corolla-tube, which is much inflated above, is 1–3 cm. long and the limb 1.5–3 cm. diameter, with broadly obovate lobes shallowly notched at the apex.

[125]

The variety *P.c. albiflos* is pure white and there is also a pale lavender-blue colour-variant. Both of these forms have a restricted geographical distribution. Although seed has been acquired on a number of occasions, the species appears to have proved intractable in cultivation.

SUBSECTION MAXIMOWICZII

The species of this subsection differ from the Eu-Nivales in being efarinose, whereas most of the Eu-Nivales are farinose. Also, with species of the subsection Maximowiczii, the corolla lobes are much longer than broad so that the petals do not present the more or less continuous rounded margin which is characteristic of the limb of the Eu-Nivales, and with species of this subsection, the corolla lobes are more or less reflexed or bent back.

P. maximowiczii Regel

This, the type species of the subsection, is a plant of northern China – Shansi and Shensi – rather than the western provinces. It is an efarinose plant, with a stout scape 20–30 cm. tall which carries a single umbel or sometimes 2–3 superposed umbels of flowers.

The species was first flowered in Britain in 1910 by Messrs. J. Veitch and Sons from seed sent back by Purdom. The colour of the corolla of the flower of cultivated plants was described as 'a clear dark purple', but collectors have reported plants with crimson and even yellow flowers, so that there may be colour-variants, some of which are reputedly dull and dingy.

The species was somewhat transient in cultivation and is not now available.

P. szechuanica Pax

This species was first discovered by Pratt, who travelled in western Szechuan in the years 1889-90. It is fairly common at an elevation of 3500–4500 m. and has since been collected on several occasions. Smith also refers to the plant as growing on the Chungtien Plateau in Yunnan, and to a form occurring in south-eastern Tibet. The latter, however, is *P. advena*.

P. szechuanica is, in common with other members of the subsection, an efarinose perennial. The stout scape is 12–50 cm. tall, at flowering time about twice the length of the leaves, and carries one, or rarely two superposed umbels of 8–15 fragrant yellow flowers.

The species was flowered in 1915 at The Royal Botanic Garden,

Edinburgh (and elsewhere) from seed sent back by Purdom. It was subsequently kept in cultivation, more or less continuously, for about twenty years – but always as the result of sowing freshly imported seed, and when seed was no longer available, the species was lost.

SUBSECTION AGLENIANA

P. agleniana Balf.f. et Forrest

This is the type species of a subsection, the three species of which are variously described by Smith and Fletcher (5) as 'truly magnificent', 'striking' and 'beautiful' respectively. The appearance of the species of the subsection in cultivation has been particularly transient and *P. elizabethae* was never flowered.

P. agleniana was first found by Forrest on the Mekong-Salween divide in 1905. Its distribution extends across north-western Yunnan, south-eastern Tibet, the Assam Himalaya and northern Burma.

The scape is 20–40 cm. high and carries a single umbel of 3–8 fragrant flowers. The corolla is more or less bell-shaped and may be as much as 4 cm. diameter at the mouth. There is a yellow eye with much farina and the tip of the petals may be slightly reflexed. The petals are usually a pale yellow, but the colour may vary from white to dark yellow and rosy pink. These colour-forms have been given the names *P.a.* var. *alba*, *P.a.* var. *atrocrocea* and *P.a.* var. *thearosa*. The two latter are found in the Assam Himalaya and Upper Burma respectively. *P.a.* var. *alba* has a wider distribution.

A NOTE ON THE CULTIVATION OF NIVALID PRIMULAS

The Nivalid primulas should be grown on a gentle slope of deep, medium-acid loam which has been made free-draining, yet moisture-retentive, by the addition of much grit and leaf-mould. During the growth period, the soil-mass should be charged with gently percolating water, and the area planted should have light to medium shade. The object is to provide that buoyant atmosphere which exists on the alps of central Asia at 5000 m.

The primula bed should be given a dusting of bonemeal twice a year. This should be pricked in and the soil surface kept free of weeds and moss.

So much for the growing season. In winter the ground should be damp, but not wet, and protection must be given to the resting buds against our freezing, icy rain. This may be done using a frame-light, a cloche, or a pane of glass. Protection may also be given by using

[127]

granulated peat as a moisture-absorptive and insulating medium, ringing the neck of the plant and covering the crown with dry peat.

But one must be careful when using peat, for it is very moisture retentive. In spring, when growth commences, if a hosepipe can be used on the primula bed then the plants will prosper. Otherwise some of the peat must be scraped away and removed, for in conditions of drought, peat will retain moisture and not release it for plant growth.

I have had some success by forming hollow, basin-like depressions in the shade of a beech hedge. The roots of the hedge-plants are pruned twice a year (with a spade) and the contours gather all applied moisture to the roots of the primulas.

These conditions can also be provided in a north-facing frame with walls of brick or concrete blocks, but not in pots.

26 THE SECTION PARRYI

This North American Section is omitted.

27 THE SECTION SIKKIMENSIS

THE SPECIES OF THE SECTION

P. alpicola	P. firmipes	*P. secundiflora*
P. chumbiensis	*P. florindae*	*P. sikkimensis*
P. didyma	*P. ioessa*	*P. waltonii*
P. erythra	*P. reticulata*	

KEY

1 Flowers yellow†	2
Flowers red, pink, mauve or purple†	6
2 Leaves heart-shaped or rounded at base	3
Leaves tapering at base	5
3 Leaves elongated	4
Leaves rounded	*florindae*
4 Leaf-stalk 1–2 times blade: blade more or less rounded at base	*alpicola*
Leaf-stalk 3–4 times blade: heart-shaped at base	*reticulata*
5 Leaf-stalk very short, conspicuously winged	*sikkimensis*
Leaf-stalk very long (1–3 times blade), slightly winged	*chumbiensis*

[128]

6 Leaves rounded or heart-shaped at base	*alpicola*
Leaves tapered or wedge-shaped at base	7
7 Leaf-stalk long at flowering time	8
Leaf-stalk nil or nearly so at flowering time	*secundiflora*
8 Corolla longer than broad, with red eye	*waltonii*
Corolla as broad as long, without red eye	*ioessa*

† There are white or near-white forms of *Pp. alpicola, ioessa, reticulata* and *sikkimensis.*

The species of this section are wholly Asiatic in distribution. They are found in the Himalaya of Nepal, Sikkim and Bhutan, in the southern and eastern provinces of Tibet, Yunnan and Szechuan in western China, and in the north-western corner of Burma.

The species from the Himalaya and China show little or no tendency to variation except in the case of *P. sikkimensis*, which is wider-ranging, occurring throughout the distribution of the section. But the species from Tibet are closely related – *Pp. alpicola, florindae, ioessa, sikkimensis* and *waltonii*. It is found that these Tibetan species will hybridize in cultivation (but not with primulas of other sections). They also hybridize in the wild, but apparently less readily, for there is presumably some habitat differential which, by preference, keeps the species apart. *P. alpicola*, however, is also known to occur in three distinct colour forms, with no suggestion of hybridization.

The species of the section vary considerably in vigour and in stature. A mature plant of one of the more vigorous species may form a number of crowns from each of which arises a scape. *P. florindae* is the giant of the section and *P. ioessa* the dwarf.

In the wild, the Sikkimensis primulas are usually associated with abundant moisture during the period of growth and flower. They grow in damp meadows and by the stream-side. If this liking for plenty of water during the growing season is catered for, then the plants are fairly easily cultivated. They could also usefully be provided with some light shade in the hotter and drier parts of the country.

These primulas are plants for the open ground. They do not require the extra care and protection during winter that is, for example, demanded by the Nivalids; and pots are anathema to the deep-thrusting, thong-like roots.

In this section, the flowers are funnel-shaped and few to many are carried in an umbel the pedicels of which may be erect or drooping, but usually erect in fruit. The foliage of the several species is distinctive.

P. alpicola Stapf (Illustration p. 30)

This perennial species develops a short, stout rootstock.

The leaves, including the petiole, are 10–30 cm. long and 2–8 cm. broad. The lamina is elliptic to oblong-elliptic, rounded at the apex and slightly cordate to shortly cuneate at the base. The margin is shallowly sharp- or round-toothed and the surface of the lamina is matt or non-shining, efarinose and without hairs, but there are numerous glands beneath. The membranously winged petiole is 1–2 times the lamina.

The scape is 15–90 cm. tall and carries one, or sometimes two superposed, umbels of numerous flowers. The pedicels are 1–8 cm. long.

The corolla is 1.5–2.5 cm. long and more or less funnel-shaped. The corolla-tube is 1–2 cm. long, which is 1–2 times the calyx. The limb is up to 3 cm. diameter, with broadly obcordate to more or less rounded lobes which are shallowly notched at the apex. Pin-eyed and thrum-eyed forms occur.

The colour of the petals is white, yellow, or violet to purple. These colour forms are fairly stable and have given rise to the varietal names *P.a. alba*, *P.a. luna* and *P.a. violacea*. Apart from the colour of the petals, there is no other morphological difference and in cultivation some hybridization occurs.

The distribution of *P. alpicola* is confined to the basin of the Tsangpo river in south-eastern Tibet. It grows in marshy ground and damp alpine pastures at an elevation of 3700–4600 m.

P. chumbiensis W.W. Sm.

This species develops a short, but comparatively stout, rhizome.

The leaves, including the petiole, are 2–12 cm. long. The lamina is 1–4 cm. long and 1–2 cm. broad, oblong or ovate-oblong to elliptic, rounded at the apex and shallowly cordate to cuneate at the base. The margin is sharp or round-toothed. The texture of the blade is leathery; it is efarinose and without hairs, slightly rough beneath and crinkled above. The distinct petiole is slightly membranously winged, and 1–3 times the lamina.

The scape is 10–25 cm. tall and carries a single umbel of 2–7 flowers. The pedicels are 1–4 cm. in flower, but as much as 8 cm. long in fruit.

The yellow corolla is 1.5–2.5 cm. long. The corolla-tube is 0.5–2 cm. long, 1–2 times the calyx. The limb is 1–1.5 cm. diameter, with more or less erect lobes which are ovate to elliptic and entire or shallowly notched at the apex. There is no annulus. Pin-eyed and thrum-eyed forms occur.

The distribution of *P. chumbiensis* is confined to the south Tibetan valley of Chumbi, between Sikkim and Bhutan, where it grows at an elevation of 5000–5300 m. In form it is a plant intermediate between *P. reticulata* and *P. sikkimensis*, but there is no suggestion of hybridization.

Only rarely available.

P. florindae Ward (Illustration p. 31)

This perennial species develops a short, stout rootstock.

The leaves, including the petiole, are 10–50 cm. long. The lamina is 4–20 cm. long and 4–15 cm. broad, broadly ovate, rounded at the apex, and deeply cordate at the base. The margin is rather irregularly sharp-toothed. Both surfaces of the lamina are without hairs. The upper surface is shining, and there are a few glands beneath. The stout petiole is 6–30 cm. long, prominently winged, and often tinged red.

The scape is up to 120 cm. tall and carries an umbel, or sometimes two superposed umbels, of up to 40 flowers, or sometimes more. The nodding pedicels are 2–10 cm. long.

The corolla is 1.5–2.5 cm. long and narrowly funnel-shaped. The colour of the petals is bright sulphur-yellow. There is no annulus. The corolla-tube is 8–15 mm. long, 1–1½ times the calyx. The limb is 1–2 cm. diameter. The lobes are broadly ovate or more or less round to slightly obovate and there is much creamy-yellow farina within the flower. The apex is slightly notched or shallowly round-toothed. Pin-eyed and thrum-eyed forms occur.

P. florindae is abundant in parts of the Tsangpo basin in southeastern Tibet, where it grows at about 4000 m. Its habitat is marshy places and even in streams and it often grows with *P. alpicola*, with which there is some hybridization. The two species may however be readily distinguished by the foliage. The leaves of *P. alpicola* are narrower and have a dull, matt surface whereas the leaves of *P. florindae* are shining and more rounded, rather like the leaves of *Caltha palustris*.

P. florindae is the giant of the section and, indeed, one of the giants of the genus – possibly rather a large plant for the small garden. But Mr. A. B. Duguid, in his garden at Coldingham in Berwickshire, has great clumps of this primula by the streamside – a magnificent plant where there is plenty of water. The species has seeded itself for about a mile downstream, and is probably the only Asiatic primula which may eventually become naturalized in suitable localities in Britain.

P. florindae is exceptional in its late and prolonged flowering period, which even in the south of England continues well into July.

P. ioessa W.W.Sm. (Illustration p. 32)

This perennial species develops a short, stout rhizome.

The leaves, including the petiole, are 6–20 cm. long and 1–2.5 cm. broad. The lamina is narrowly oblong or oblanceolate to spathulate, rounded to more or less acute at the apex and tapered at the base. The margin is saw-toothed to deeply cut sharp-toothed. The lamina is of a papery texture, efarinose and without hairs. The slightly winged petiole is shorter than or more or less equal to the lamina.

The scape is 10–30 cm. tall and carries an umbel of 2–8 flowers, although exceptionally a good head may have up to 15 flowers fairly closely spaced. The slender pedicels are 1–6 cm. long.

The funnel-shaped corolla is 2–2.5 cm. in length and diameter. The colour of the petals is pale-madder pink or pinkish-mauve or white and there are particularly good forms which are deep violet and of a delightful fragrance. The corolla-tube is 6–20 mm., 1–2 times the calyx, and quickly widens into the limb. The lobes are broadly obcordate and more or less entire or shallowly notched at the apex. Pin-eyed and thrum-eyed forms occur.

P. ioessa grows in wet meadows at upwards of 3500 m. in southeastern Tibet. In cultivation the species sets viable seed, from which it may be raised. Good forms may also be divided after flowering. The need for plenty of water during the growing season must be catered for and a little shade helps in this respect.

P. reticulata Wall.

P. reticulata is a species of very variable stature which usually develops a short rhizome.

The leaves, including the petiole, are 4–40 cm. long. The lamina is 2–10 cm. long, oblong to ovate-oblong, rounded at the apex and cordate at the base. The upper surface of the lamina is smooth or slightly wrinkled. The lower surface is without hairs or faintly, shortly and softly hairy. The petiole is up to 30 cm. long, 3–4 times the lamina.

The scape is 18–45 cm. tall, slender at first, but later more robust, and carries a head of few to many flowers, the outer generally nodding, the inner more or less erect. The slender pedicels are 1–5 cm.

The corolla is yellow or white. There is no annulus. The corolla-tube is 5–25 mm. long, 1–2 times the calyx. The limb is 1–2 cm. diameter, with obovate lobes very shallowly notched, or wavy, at the apex. Pin-eyed and thrum-eyed forms occur.

This close ally of *P. sikkimensis* grows at 3500–5000 m. in Nepal, Sikkim, southern Tibet and western Bhutan. In cultivation it is rarely available, but viable seed has been offered in India.

P. secundiflora Franch. (Illustration p. 33)

This species is a usually sturdy perennial which develops a short, stout rhizome.

The leaves, including the petiole, are 3–30 cm. long and 1–4 cm. broad. The lamina is oblong or oblong-elliptic to obovate or oblanceolate; obtuse, rounded or acute at the apex and at the base tapering into the prominently winged petiole which is usually very short, but may equal the lamina in autumn. The margin is finely round-toothed to deeply saw-toothed. The lower surface is yellow-farinose when young. Both surfaces are without hairs.

The stout scape is 10–90 cm. tall and carries an umbel, or rarely two superposed umbels, of 5–20 nodding flowers in a more or less one-sided (secund) inflorescence, whence the specific epithet. The pedicels are 1–5 cm. long, slender in flower, but stiff and erect in fruit.

The funnel-shaped corolla is 1.5–2.5 cm. long. The colour of the petals is reddish-purple or deep rose-red. There is no annulus. The limb is 1.5–2.5 cm. diameter, with obovate-oblong lobes which are rounded or slightly retuse at the apex. Both pin-eyed and thrum-eyed forms occur.

P. secundiflora is a plant of wet meadows and pastures at 3500–4000 m. in south-western Szechuan and north-western Yunnan. It is a beautiful plant and ought to be grown more frequently than appears to be the case.

[133]

P. sikkimensis Hook. (Plates I and XXIII)

This species is a sturdy perennial which develops a short, stout rhizome.

The leaves, including the petiole, are 10–40 cm. long and 2–7 cm. broad. The lamina is elliptic or oblong to oblanceolate, rounded or occasionally acute at the apex, usually long-tapered at the base, but sometimes obtuse or even more or less rounded. The margin is round-toothed, saw-toothed or sharp-toothed. The lamina is shining, without hairs and efarinose, but with a few glands below. The petiole is much shorter than the lamina, conspicuously winged and not clearly distinguished from the lamina, or more or less equal to the lamina and then quite distinct.

The scape is 15–90 cm. tall, thickening in fruit, and carries one umbel or sometimes two superposed umbels, of numerous flowers. The pedicels are 2–10 cm. long, pendulous at first, but a little longer and erect in fruit.

The corolla is 2–3 cm. long. The colour of the petals is yellow or occasionally creamy-white. There is no annulus. The corolla-tube is 7–24 mm. long, 1–2 times the calyx. The limb is 1.5–3 cm. diameter with oblong to obcordate lobes which may be entire or shallowly notched at the apex. Pin-eyed and thrum-eyed forms occur.

This, the type and most widespread species of the section, is a gregarious plant of wet meadows at 4000–5500 m. Its distribution includes the Himalaya of Nepal, Sikkim and Bhutan, southern and eastern Tibet, the Chinese provinces of Szechuan and Yunnan, and the extreme north-west of Burma.

As may be expected of a plant with wide geographical and altitudinal distribution, this species can be rather variable, and various forms have become more or less locally stabilized. Thus *P. s.* var. *hopeana* may be described as a slender form of *P. sikkimensis* from Bhutan and southern Tibet at 4500–5000 m.; the corolla is white or pale yellow fading to white. *P.s.* var. *pudibunda* is an extreme alpine form, with more or less the distribution of the species.

P. waltonii Watt (Illustration p. 34)

This species is a perennial with a short, stout rootstock.

The leaves, including the petiole, are 8–30 cm. long and 2–7 cm. broad. The lamina is elliptic-oblong to oblanceolate, rounded at the apex and cuneate at the base. The margin is raggedly sharp- or round-toothed to sharply saw-toothed. The lamina is efarinose and both surfaces are without hairs.

The scape is 20–70 cm. tall and carries an umbel of few to many flowers. The pedicels are 1–7 cm. long, more or less nodding in flower, but elongating slightly and becoming erect in fruit.

The narrowly or broadly funnel-shaped corolla is 1.5–3 cm. long. The colour of the petals is pink to deep wine-purple or dull dark lilac. There is no annulus. The corolla tube is 0.5–2 cm. long, 1–3 times the calyx. The limb is 0.5–2 cm. diameter with lobes entire or shallowly notched at the apex. The lobes are powdered inside with white or yellowish farina, except at the centre, where there is a rosy eye. Pin-eyed and thrum-eyed forms occur.

P. waltonii has the reputation of not being a good doer, and it is doubtful if the true species remains in cultivation. Plants are, however, offered which are described as *P. waltonii* hybrids. Many of these are nice little plants and better growers. The colour of the corolla often suggests crossing with yellow species of the section.

The distribution of *P. waltonii* is south-eastern Tibet and Bhutan.

28 THE SECTION CANDELABRA

THE SPECIES OF THE SECTION

*P. anisodora	*P. helodoxa	*P. poissonii
*P. aurantiaca	*P. ianthina	P. polonensis
*P. beesiana	P. imperialis	P. prenantha
P. brachystoma	*P. japonica	*P. prolifera
*P. bulleyana	P. khasiana	*P. pulverulenta
*P. burmanica	P. mallophylla	*P. serratifolia
P. chrysochlora	P. melanodonta	*P. smithiana
*P. chungensis	P. microloma	P. stenodonta
*P. cockburniana	P. miyabeana	P. sumatrana
P. cooperi	P. morsheadiana	*P. wilsonii

KEY

1 Flowers red to purple†	2
Flowers yellow or orange	9
2 Farinose inside calyx or elsewhere. Seed-capsules globular	3
Completely efarinose. Seed-capsules oblong	7

3 Farina on inside of calyx only	4
Farina on scape and pedicels also	5
4 Corolla-tube 3 times calyx	*japonica*
Corolla-tube 2 times calyx	*burmanica*
5 Farina sulphur-coloured	*ianthina*
Farina white	6
6 Corolla-tube and eye yellow	*beesiana*
Corolla-tube and eye red or purple	*pulverulenta*
7 Corolla-limb flat, 2–3 cm. across; petals deeply notched	*poissonii*
Corolla-tube concave, 1.5 cm. across; petals scarcely notched	8
8 Flowers pale purple to red, with yellow eye	*wilsonii*
Flowers dark blackish purple, with greenish eye	*anisodora*
9 Farina on scape, pedicels and calyx	10
Completely efarinose	14
10 Calyx-segments awl-shaped. Midrib of leaves red	*bulleyana*
Calyx-segments triangular. Midrib of leaves not red	11
11 Flowers yellow	12
Flowers orange	13
12 Corolla-tube 2 times calyx	*smithiana*
Corolla-tube 3 times calyx	*helodoxa*
13 Dark reddish orange. Corolla-tube not more than 2 times calyx	*cockburniana*
Pale orange. Corolla-tube 3 times calyx	*chungensis*
14 Bracts at least as long as pedicels. Flowers deep reddish orange	*aurantiaca*
Bracts shorter than pedicels. Flowers yellow at least in part	15
15 Flowers clear yellow	*prolifera*
Flowers yellow with orange eye and stripes .	*serratifolia*

† P. *beesiana* var. *leucantha* is white.

The species of the section Candelabra are all perennials. Most are robust plants, although a very few are dwarf and short-lived.

The root is a short, thick rhizome, on which, as summer progresses, develop axillary buds. From each of these springs a whorl of leaves, to produce the multiple-crown plants of autumn. In early spring, when growth is first commencing, these axillary buds may be teazed apart and replanted in open ground. With careful cultivation, they may flower later in the year. The scape, of course, is terminal, and dies back after the ripening of the seed.

Most, but not all, species of the section are entirely herbaceous, but *P. poissonii* and its allies over-winter as a number of green rosettes, with much reduced leaves, and *P. cockburniana* differs from all other species of the section. With the ripening of seed in autumn, the rosette of leaves of *P. cockburniana* rots to the base and exposes the axillary renewal buds. These are arranged in a circle round the base of the dying scape. The buds develop thick, fleshy scale leaves, and root while still attached to the plant. This therefore provides one method of propagation. The green scale leaves can also be propagated in the same way as the scale-leaves of a lily or tulip.

The scape of species of the section is quite distinctive. The flowers are carried in a series of rather loose, superposed whorls, or sometimes only a single whorl, and hang down at the ends of the more or less horizontal pedicels.

The distribution of the section is entirely Asiatic, the main concentration of species being south-western China, particularly the provinces of Yunnan and Szechuan. Some species extend into alpine Burma and south-eastern Tibet. The most westerly extension in the Himalaya is *P. ianthina* from Sikkim, and there is one species each from Japan, Taiwan and Java.

The Candelabra primulas are mostly vigorous plants, plants for the open garden and not the alpine house. They are quite amenable to cultivation, and do not require excessive precautions against rotting of the collar and temperature fluctuation. In the wild they are plants of marshy places and therefore like plenty of water during the period of growth. They are, however, much more drought-resistant than, for example, the Nivalid primulas.

With primulas of this section it might be suggested that the flower-to-foliage ratio is low, for they are large-leaved plants, particularly during and after flowering. But while this might hold for a single plant, they look very fine as a group by the stream side, or among dwarf shrubs.

P. anisodora Balf.f. et Forrest (Plate XXVIII)

The leaves, including the petiole, are 25 cm. long and 8 cm. broad. The lamina is obovate, obtuse or rounded at the apex and tapering at the base into the broad, winged petiole. The lamina is efarinose and without hairs and the margin has diminutive sharp teeth.

The scape is up to 6 cm. tall and carries 3–5 rather remote, super-posed umbels each of which has 8–10 flowers. The pedicels are 1–1.5 cm. long, half-nodding in flower but erect in fruit.

The funnel-shaped corolla is dark crimson or blackish purple and there is a thin annular ring at the greenish throat. The corolla-tube is up to 1 cm. long, about 2 times the calyx. The concave limb is up to 1.5 cm. diameter. The overlapping lobes are more or less qua-drangular, slightly wavy at the margin and may be shallowly notched at the apex. Both pin-eyed and thrum-eyed forms occur.

P. anisodora comes from Yunnan and Szechuan in western China, where it grows at 3500–4000 m. in moist pastures.

The specific epithet derives from the fragrance of the foliage which, particularly in early spring, is reminiscent of aniseed. In cultivation, *P. anisodora* crosses readily with *P. helodoxa*, the flowers of the hybrid being of a crushed-strawberry shade.

P. aurantiaca W.W.Sm. et Forrest

The leaves, including the petiole, are up to 20 cm. long and 5 cm. broad. The lamina is oblanceolate to obovate, rounded at the apex and tapering at the base, usually rapidly. The lamina is raggedly sharp-toothed at the margin, efarinose, without hairs on the upper surface but with a slight dusting of powder beneath.

From the rosette of leaves arise one or two scapes up to 30 cm. tall. These are tinged red and are without hairs and carry 2–6 whorls, each of 6–12 flowers. The reddish pedicels are up to 1 cm. long.

The corolla is deep reddish orange and there is a faint annulus. The corolla-tube is 8–16 mm. long, 1–2 times the calyx. The limb is 1 cm. diameter, with narrowly obovate to more or less oblong lobes which are slightly notched at the apex. Pin-eyed and thrum-eyed forms occur.

P. aurantiaca comes from Yunnan where it grows, at an altitude of about 3500 m., beside streams in moist alpine pastures.

P. beesiana Forrest

The leaves, including the petiole, are up to 22 cm. long and 6 cm. broad, but may exceed 40 cm. in fruit. The lamina is ovate-lanceolate to obovate, rounded at the apex and sharply tapered at the base. The margin is shallowly sharp-toothed and there are potentially farinose glands on the lower surface of the lamina.

The scapes are stout and there is usually more than one scape to each plant. Each scape carries 2–8 whorls of flowers, with 8–16 flowers in each whorl. The pedicels are 1–3 cm. long.

The corolla-tube is orange and 1–1.5 cm. long, about 2 times the calyx. The limb is about 2 cm. diameter, rose-carmine with a yellow eye. The obovate lobes are deeply notched at the apex. Both pin-eyed and thrum-eyed forms occur.

P. beesiana is closely related to *P. pulverulenta*, for both species have reddish or purplish flowers and farina on the nodes of the scape, the pedicels and the calyx. But *P. beesiana* has ovate-lanceolate leaves which are broadest at or near the middle, while *P. pulverulenta* has obovate to oblanceolate leaves which are broadest near the apex. Further, *P. beesiana* has a yellowish corolla-tube and a yellow eye while *P. pulverulenta* has a red or purple corolla-tube and eye.

P. beesiana is a plant of Yunnan and Szechuan, where it grows at an altitude of about 2600 m. in moist mountain meadows.

P. bulleyana Forrest

The leaves, including the petiole, are 12–35 cm. long and 3–10 cm. broad. The midrib is red. The lamina is ovate to ovate-lanceolate, rounded at the apex and tapering rapidly at the base. The margin is irregularly sharp-toothed and there are a few potentially farinose glands on the lower surface of the lamina.

The stout scape is up to 70 cm. tall and carries 5–7 whorls of flowers. There is a thin covering of farina on the pedicels, which are up to 3 cm. long and almost erect in fruit.

In bud the corolla is a deep crimson, but deep orange when mature. The corolla-tube is about 1.6 cm. long, 2 times the calyx, and the annulate limb is 2 cm. diameter with broadly obovate lobes shallowly notched at the apex. Both pin-eyed and thrum-eyed forms occur.

P. bulleyana comes from Yunnan where it grows in a similar type of habitat to *P. beesiana* : marshes, mountain meadows, ditches and streams – but at the slightly higher altitude of 2900–3200 m. The two species should not however be confused, for *P. bulleyana* has orange-yellow flowers and a red petiole and midrib while *P. beesiana* has rose-carmine or deep red flowers and a lilac or flesh-coloured petiole and midrib.

P. burmanica Balf.f. et Ward

The leaves, including the petiole, are up to 30 cm. long and 8 cm. broad. The lamina is oblanceolate, rounded at the apex and tapered into the winged petiole at the base. The margin is raggedly sharp-toothed and both surfaces of the lamina are efarinose and without hairs, although there are a few potentially farinose glands on the lower surface.

The scape is as much as 60 cm. tall and carries up to 6 whorls of flowers, there being 10–18 flowers in each whorl. The pedicels are about 2 cm. long.

The corolla is reddish-purple with a yellow eye. The corolla-tube is about 2 cm. long, 2 times the calyx, and the limb is 2 cm. diameter, with obcordate lobes which are entire, shallowly notched at the apex, or with diminutive rounded teeth.

P. burmanica is a plant of Yunnan and Upper Burma, where it grows in lower-altitude meadows and in wet forests. It is closely allied to *P. beesiana* from Yunnan, but the two species are readily separated because *P. burmanica* is completely efarinose except for the inside of the calyx lobes whereas *P. beesiana* has copious farina on both surfaces of the calyx and on the pedicels and scape.

The descriptions given are those of the true species. In the wild these are kept apart by geographical distribution and altitudinal range. But in cultivation there is a strong tendency to hybridize between *Pp. aurantiaca, beesiana, bulleyana, burmanica* and *chungensis* and plants of intermediate form often occur.

P. burmanica grows well in a rich loam, provided plenty of water is given during the growing season.

P. chungensis Balf.f. et Ward

The leaves, including the petiole, are 10–30 cm. long and 3–10 cm. broad. The lamina is elliptic to oblong or oblong-obovate, rounded at the apex and cuneate or more sharply tapered at the base. The margin is obscurely and shallowly lobed and irregularly sharp-toothed. Both surfaces of the lamina are efarinose and without hairs, although there is often a very thin scattering of potentially farinose glands.

The stout scape, farinose at the nodes, is up to 80 cm. tall and carries from 2–5 whorls of flowers, with as many as twelve flowers to each whorl. The pedicels are about 2 cm. long in flower, but 3 cm. in fruit.

The annulate corolla is pale orange. The corolla-tube is 1.5 cm. long, 3 times the calyx, and the limb is 1.5–2 cm. diameter, with broadly obovate lobes which are shallowly notched at the apex.

The distribution of *P. chungensis* includes Szechuan, Yunnan, the

Assam frontier and Bhutan, where it grows at an altitude of 2900–3200 m. in marshy places in the forest.

In cultivation *P. chungensis* likes an open situation and its roots in a sandy soil charged with plenty of leafmould, and plenty of water during the growing season.

P. cockburniana Hemsl. (Illustration p. 35)

The leaves, including the petiole, are up to 15 cm. long and 4 cm. broad. The lamina is oblong to oblong-obovate, rounded at the apex, and round and then suddenly cuneate or more sharply tapered, at the base. The margin is obscurely and shallowly lobed and minutely sharp-toothed. Both surfaces of the lamina are efarinose and without hairs, but there is a thin scattering of potentially farinose glands below.

The slender scape, farinose at the nodes, is up to 40 cm. tall and carries two or three superposed whorls of flowers, or sometimes only one whorl. The farinose pedicels are up to 3 cm. long.

The annulate corolla is dark orange tinged red. The corolla-tube is about 1 cm. long, 1½–2 times the calyx. The limb is 1.5 cm. diameter, with oblong-obovate lobes. Flowers of this species develop only the pin-eyed form, the style reaching almost to the annulus.

P. cockburniana appears to be confined to south-western Szechuan, where it grows in marshy alpine meadows at an altitude of 2900–3200 m.

There is an obvious relation between *Pp. cockburniana* and *chungensis*. The latter species, however, is much the more vigorous plant and has much paler orange flowers, which lack the red tint of *P. cockburniana*. Further, the leaves of *P. cockburniana* are smaller than those of *P. chungensis*, and they are more distinctly petioled and have the margin regularly but shallowly sharp-toothed, instead of irregularly sharp-toothed.

P. cockburniana is usually treated as a biennial. It likes a half shady, moist but well drained situation in good, light loam.

P. helodoxa Balf.f. (Plate XXIX)

The leaves, including the petiole, are up to 35 cm. long and 7 cm. broad. The lamina is oblong-obovate or lanceolate, obtuse or rounded at the apex and tapering rapidly at the base into the usually long, winged petiole. The margin is finely sharp-toothed and both surfaces are efarinose and without hairs or sometimes faintly farinose below.

The stout scape, farinose at the nodes, is 100–120 cm. tall and carries 4–6 whorls of flowers with up to 20 flowers in each whorl. The farinose pedicels are 1–2.5 cm. long.

The annulate corolla is bright golden yellow and fragrant. The corolla-tube is 1.5–2 cm. long, about 3 times the calyx. The limb is 2.5 cm. diameter, with obovate lobes deeply notched at the apex. Both pin-eyed and thrum-eyed forms occur.

P. helodoxa is a native of the extreme west of Yunnan and has also been found in north-eastern Burma. It grows at the low altitude of 2000 m. in open situations by streams, in marshy meadows and on clayey pasture land.

P. ianthina Balf.f. et Cave

The leaves, including the petiole, are up to 25 cm. long and 5 cm. broad. The lamina is oblong-oblanceolate, rounded at the apex and tapering rapidly at the base into the winged petiole. The margin has small, sharp teeth and there are pitted glandular depressions on both surfaces of the lamina, and potentially farinose glands beneath.

The stout scape is up to 60 cm. tall, sulphur-farinose at the nodes, and carries 3 whorls of flowers with about 12 flowers to each whorl. The slightly farinose pedicels are about 1 cm. long, but elongate to 2 cm. in fruit.

The corolla is violet. The corolla-tube is about 1 cm. long, 2 times the calyx. The limb is 1.5–2 cm. diameter with obcordate or more or less rounded lobes, which are nearly 1 cm. diameter and shallowly notched at the apex.

P. ianthina is a Sikkim plant. It has always been rare, but viable seed has been offered in India.

P. japonica A. Gray

The leaves, including the petiole, are up to 25 cm. long and 8 cm. broad. The lamina is obovate-oblong to broadly spathulate, rounded at the apex and tapering sharply at the base into the winged petiole, or sometimes more or less sessile. The margin is finely and irregularly round- or sharp-toothed.

The stout scape rises to a height of 45 cm. and carries 1–6 whorls of flowers. These are borne on pedicels 2 cm. long, which elongate to 3 cm. or more in fruit.

The corolla is usually purplish red. The corolla-tube is 2–2.5 cm. long, 3 times the calyx, and the limb is 2 cm. diameter, with obcordate lobes. The flowers of this species are monomorphic: only thrum-eyed forms develop.

[142]

As the specific epithet implies, *P. japonica* is a Japanese plant. It is easily cultivated, and likes a moist, shady situation in a rich, rather heavy, and preferably lime-free loam. The species is a perennial, but better flowers are obtained by frequently raising from seed. And while many colour-forms of this species are poor, there are also some particularly good ones (including a good white), a situation which demands selective cultivation.

P. poissonii Franch.

The leaves, including the petiole, are up to 18 cm. long and 4 cm. broad. The lamina is oblong-obovate, rounded at the apex and tapering rapidly at the base into the petiole which may be scarcely differentiated or conspicuously winged. Even at maturity, the lamina tends to be somewhat rolled inwards and the margin is regularly and finely sharp-toothed.

The scape is up to 45 cm. tall and carries 2–6 whorls of flowers. The pedicels are 1–2 cm. long and, especially in fruit, closely adpressed to the scape.

The distinctly annulate corolla is deep purplish crimson with a yellow eye, or sometimes white. The corolla-tube is 1–1.5 cm. long, 2–3 times the calyx. The spreading limb is 2–3 cm. diameter, with obcordate lobes deeply notched at the apex, or sometimes round toothed.

P. poissonii is a plant of western Szechuan and western Yunnan, where it grows at 2800–3000 m. The two closely allied species *Pp. anisodora* and *wilsonii* are aromatic, whereas *P. poissonii* is completely efarinose and not aromatic.

P. prolifera Wall.

The leaves, including the petiole, are up to 30 cm. long and 6 cm. broad. The lamina is oblong or more or less spathulate, rounded or obtuse at the apex and tapering rapidly at the base. The margin is minutely sharp-toothed. The leaves are efarinose, and marked with glandular dots below.

The scape is up to 60 cm. tall, about 2 times the leaves, and carries 1–3 whorls of flowers. The pedicels are 1–3 cm. long and slightly nodding in flower, but erect in fruit.

The annulate corolla is yellow. The corolla-tube is 1–1.5 cm. long, 2–3 times the calyx. The salver- or bowl-shaped limb is 2 cm. diameter, with obcordate lobes which are round-toothed and shal-

lowly notched at the apex. Pin-eyed and thrum-eyed forms occur.
P. prolifera is a plant from the Khasia mountains of Assam.

P. pulverulenta Duthie (Illustration p. 36)

The leaves, including the petiole, are up to 30 cm. long and 10 cm.
broad. The lamina is obovate or oblanceolate, rounded at the apex
and tapering at the base into the winged petiole. The margin is
regularly or irregularly sharp-toothed or sometimes obscurely lobed.
There is a thin scattering of potentially farinose glands on the lower
surface of the leaf blade.

The scape is up to 100 cm. tall and carries several superposed
whorls of flowers. The spreading pedicels are about 2 cm. long in
flower, but longer in fruit. The scape, pedicels and calyx are all
covered with white farina.

The annulate corolla is red with a darker red or purple eye. The
corolla tube is about 1.5 cm. long, about 2 times the calyx. The limb
is 2–3 cm. diameter, with obcordate lobes deeply notched at the apex.
Pin-eyed and thrum-eyed forms occur.

P. pulverulenta comes from the hills of western Szechuan.

In cultivation, the species likes a good, rich loam in a damp, open,
or half shady position, preferably on the banks of a pond or stream.

The colour of the corolla varies considerably and the better forms
have been identified as various strains: the Lapworth, Hidcote and
Bartley strains for instance.

P. serratifolia Franch.

The leaves, including the petiole, are up to 20 cm. long and 5 cm.
broad. The lamina is ovate-oblong to obovate, rounded at the apex
and tapering at the base. The margin is irregularly and deeply sharp-
toothed, or raggedly sharp-toothed, and the plant is completely
efarinose.

From each plant arise one or two stout scapes which are up to
45 cm. tall and carry a single terminal umbel, or two superposed
umbels, with 5–10 flowers in each umbel. The pedicels are 1–2 cm.
long and slightly nodding at first, but 3 cm. long and upright in fruit.

The annulate corolla is yellow and each petal has a distinct bar of a
deeper orange-yellow running from the mouth of the tube to the
tip of the lobe. The corolla-tube is 1–1.5 cm. long, 2 times the calyx,
and the limb is 2.5 cm. diameter with obovate lobes which may be
entire or shallowly notched at the apex. Both pin-eyed and thrum-
eyed forms occur.

The distribution of *P. serratifolia* includes Yunnan, south-eastern Tibet and Upper Burma. It has always been a very rare plant in cultivation.

P. smithiana Craib.

The leaves, including the petiole, are up to 20 cm. long and 5 cm. broad. The lamina is oblong-lanceolate to oblanceolate, rounded at the apex and tapered at the base into the petiole which is more or less distinct and conspicuously winged. Both surfaces of the lamina are efarinose and without hairs, or thinly powdered with yellowish farina beneath. The margin is regularly or irregularly shallowly sharp-toothed.

From the rosette of leaves arise 1–3 stout scapes up to 60 cm. tall. Each scape carries 1 or, more usually, 2–4 superposed whorls of flowers. The pedicels are about 2 cm. long.

The annulate corolla is pale yellow. The corolla-tube is about 1 cm. long, 2 times the calyx. The limb is 1–2 cm. diameter with lobes oblong to oblong-obovate and shallowly notched at the apex.

P. smithiana is a plant of the eastern Himalaya, its distribution extending from central Bhutan to the Chumbi valley, where it grows at an altitude of 2900–3500 m. In cultivation it likes an open, sunny situation in good, rich fibrous loam, and it must be provided with plenty of water during the growing season.

P. wilsonii Dunn

The leaves, including the petiole, are up to 20 cm. long and 5 cm. broad. The lamina is oblanceolate, rounded at the apex, and at the base tapering sharply into the conspicuously winged petiole. The lamina is finely sharp-toothed at the margin, which tends to be slightly recurved. *P. wilsonii* is closely related to *P. poissonii*. Both species have bluish-green leaves, but *P. wilsonii* less so, and both species retain leaves in winter.

The stout scape is about 90 cm. tall and carries 3–6 whorls of flowers. The pedicels are about 2 cm. long.

The annulate corolla is purple. The corolla-tube is 1.5–2 cm. long, 2½–3 times the calyx. The concave limb is about 1.5 cm. diameter. The lobes are more or less round, usually longer than broad, and entire, shallowly round-toothed or shallowly notched at the apex.

P. wilsonii is a plant of the mountains of western Szechuan and western Yunnan.

29 THE SECTION AURICULA

This predominantly European Section is omitted.

30 THE SECTION FLORIBUNDAE

The Species of the Section

P. *aucheri*   P. *gaubaeana*   P. *verticillata*
P. *floribunda*

The species of this section have the following characteristics. They all have yellow flowers. The corolla-tube is usually narrow-cylindrical, long in comparison with the proportion of the flower, and 2–3 times the calyx. The limb is of small diameter.

The flowers are carried in a single umbel, or in several superposed whorls.

The margin of the leaf is involute: it is rolled upwards, which places the section adjacent to Auricula.

These species mostly grow on the banks of streams in dry arid areas. They are all Asiatic, but distribution is widely separated and discontinuous. P. *aucheri* is from Muscat in eastern Arabia and P. *floribunda* from Kumaon in the western Himalaya. The altitudinal range of the latter species is 500–2000 m., which is low for the genus in the Himalaya. P. *floribunda* therefore, in common with all the other species of the section, is not hardy. The remaining two species, Pp. *gaubaeana* and *verticillata*, come from Luristan in south-western Persia and the Yemen in Arabia, respectively.

P. × *kewensis*. This plant was the result of a natural cross between Pp. *verticillata* and *floribunda* which occurred at Kew. The hybrid originally proved to be sterile, but subsequently gave rise to a fertile, tetraploid form. This is a floriferous plant of vigorous constitution and is often grown as a house plant. It usually breeds true from seed. The name P. × *kewensis* is now applied to this latter fertile form, and not the original cross.

APPENDIX
SOME SOURCES OF THE ASIATIC
SPECIES OF PRIMULA

The chief source of plants of the Asiatic species of primula is the nurserymen who advertise in the pages of the Society's Bulletin, and a few others who do not so advertise. A few nurserymen also have a useful seed list.

But by no means all nurseries offer Asiatic primulas. They are plants which are grown particularly well in some districts, but only with difficulty in others, and it is a matter of patient enquiry and search before one finds the real primula growers.

There are also the various seed exchanges, notably:

The Alpine Garden Society, Secretary: E. M. Upward, Lye End Link, St. John's, Woking, Surrey.
The Scottish Rock Garden Club, Hon. Secretary: Mrs. I. J. Simpson, 48 St. Alban's Road, Edinburgh EH9 2LU.

Plants grown from seed, particularly exchange seed, should always be checked for identity. Very occasionally, plants supplied by a nurseryman may also be incorrectly named.

Two Indian firms from whom seed may be obtained are:

1 G. Ghose & Co., Townend, Darjeeling, India.
2 P. Kohli & Son, Park Road, Srinagar, Kashmir, India.

Both of these firms will accept payment in crossed British postal orders.

REFERENCES

1 Holdsworth, R. L. *Plant Hunting on the Kashmir-North West Frontier Border*. A.G.S. Bulletin Vol. 8 (1940) p. 27.

2 Pax F. and Knuth R. in Engler's *Das Pflanzenreich*. Heft 22, Primulaceae. Leipzig 1905.

3 Balfour I. B. *Chinese Species of Primula*. Journ. R.H.S. Vol. 39 (1913) p. 130.

4 Smith W. W. and Forrest G. *The Sections of the Genus Primula*. Journ. R.H.S. Vol. 54 Part 1 (1929) p. 4.

5 Smith, Sir W. W. and Fletcher, Dr. H. R. *Various papers*. Trans. Roy. Soc. Edin. Vols. LX and LXI, 1941–8.

6 Idem. *Various papers*. Trans. Bot. Soc. Edin. Vols. XXXIII and XXXIV, 1941–7.

7 Smith, Sir William Wright and Fletcher, Dr. H. R. *The Section Soldanelloideae of the genus Primula*, Journ. Lin. Soc. Bot. Vol. 52 (1942) pp. 321–35.

8 Cox, E. H. M. *Plant Hunting in China*. Collins. London 1945.

9 Fletcher H. R. *A Quest of Flowers*. Edinburgh U. P. 1975.

10 Saunders, Mrs. C. B. *A note on Primula Root Aphis*. A.G.S. Bulletin Vol. 5 (1937) pp. 124–9.

11 Masterton, R. S. *Propagation by Leaf-Cuttings of the Petiolaris Primulas*. Journ. S.R.G.C. No. 9 (1951) p. 20.

12 Livingstone, D. *Primula bracteosa*. A.G.S. Bulletin Vol. 14 (1946) pp. 59–60.

REFERENCES OF THE SECTIONS OF THE GENUS PRIMULA
IN SMITH & FLETCHER (5), (6) & (7)

It has been mentioned that this, the standard work on the Primula, is distributed throughout the proceedings of three learned societies (5) (6) and (7). The following list may therefore be of assistance to anybody wishing to study the morphology of a species in greater detail, or to study a species not described in this handbook. The very few species which have been found since the standard work was published are usually described in the Journal of the Royal Horticultural Society. The description can be located by reference to Index Kewensis.

GLOSSARY

Note: In this simplified list the words are defined only in relation to their use *in this book*. This does not in all cases cover the full range of their meanings.

acute Pointed.
annulus A thickened ring just inside the throat of the flower.
annulate Having an annulus.
attenuate (of leaf-blade) Tapering gradually into the leaf-stalk.
bullate Blistered or puckered.
campanulate Bell-shaped.
capsule Seed-pod.
cordate Heart-shaped.
crenate With rounded teeth.
crenulate With tiny rounded teeth.
cuneate Wedge-shaped.
dentate Sharp-toothed.
depressed (of flowers) Sharply down-turned.
dimorphic In two forms (e.g. pin-eyed and thrum-eyed).
efarinose Having no farina.
entire (of leaf-margins) Smooth, undivided.
farina Flour-like powder (usually white or yellow) on leaves, stems, etc.
farinose Having farina.
glabrous Hairless, usually smooth.
glaucous (of leaves) Having a whitish bloom, resulting in a bluish or greyish green.
globose Spherical, rounded.
herbaceous Soft-leaved, non-woody, dying to the ground each year.
inflorescence The whole flower cluster, including the flower-stalks.
involute Margins of young emerging leaves rolled inwards, towards the upper side.
lamina Leaf-blade.
lanceolate Lance-shaped.
limb (of a flower) The expanded part of the corolla beyond the tube.
linear Long, narrow, with parallel sides.
monomorphic Of one form (as opposed to dimorphic).
obcordate Reverse heart-shaped (the notch at the apex).
oblanceolate Lanceolate but with the broadest part nearer the apex.
obovate Egg-shaped with the broader end nearer the apex.
obtuse (of leaf, petal, etc.) Blunt-ended.
orbicular More or less circular in outline.

ovate Egg-shaped, broadest nearer the base.

palmate (of a leaf) With divisions and veins radiating from a common point.

pedicel The stalk of a single flower.

pedicellate Having pedicels.

pendant Hanging, nodding.

petiole Leaf-stalk.

pin-eyed With the pistil, not the stamens, showing at the throat of the flower.

pinnate (of leaf) Divided into separate leaflets arranged along the two sides of the leaf-stalk (feather-like).

pinnatifid Like pinnate but the division not reaching down to the mid-vein.

raceme A flower-head with the flowers in a loose spike on pedicels of roughly equal length, the lowest opening first.

reflexed (of flowers) Bent abruptly downwards.

reniform Kidney-shaped.

retuse With rounded apex having a shallow notch.

revolute Margins of young emerging leaves rolled outwards, towards the lower side.

rhizome The base of the stem, underground, at its junction with the roots; often thickened or creeping.

rootstock = *rhizome*.

sagittate Arrowhead-shaped.

scape The leafless main stalk of a flower-head.

serrate Saw-toothed, the sharp teeth pointing forward.

sessile Stemless.

spathulate (*spatulate*) Spoon- or paddle-shaped.

spicate (of a flower-head) In a close spike, the individual flowers sessile.

stolon A stem, on or under the ground, rooting at the tip.

stoloniferous Having stolons.

subcordate Somewhat heart-shaped.

suborbicular Somewhat circular.

subrotundate Somewhat rounded in outline.

superposed (of umbels) One above the other, on the same stem.

thrum-eyed With the cluster of stamens, not the pistil, showing at the throat of the flower.

truncate (of base of leaf-blade) Cut-off squarely.

umbel A flat or rounded flower-cluster with the flower-stalks arising from a common point.

undulate (of edge of leaf) Wavy.

venation The arrangement of the veins.

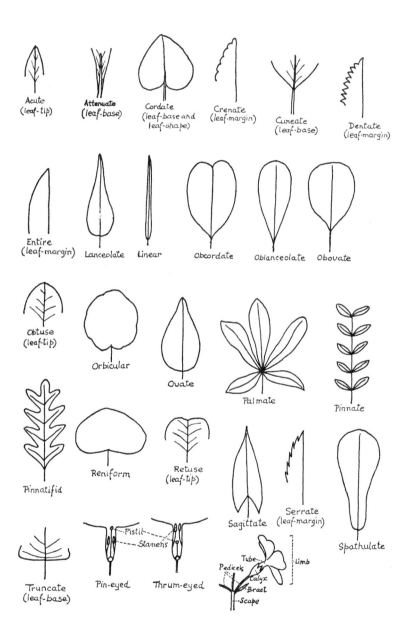

Acute
(leaf-tip)

Attenuate
(leaf-base)

Cordate
(leaf-base and
leaf-shape)

Crenate
(leaf-margin)

Cuneate
(leaf-base)

Dentate
(leaf-margin)

Entire
(leaf-margin)

Lanceolate

Linear

Obcordate

Oblanceolate

Obovate

Obtuse
(leaf-tip)

Orbicular

Ovate

Palmate

Pinnate

Pinnatifid

Reniform

Retuse
(leaf-tip)

Sagittate

Serrate
(leaf-margin)

Spathulate

Truncate
(leaf-base)

Pin-eyed

Pistil

Stamens

Thrum-eyed

Tube
Pedicels
Calyx
Bract
Scape

Limb

INDEX AND SYNONYMS

This Index aims to be comprehensive within the geographical limits of the book. That is to say, all names known to the author to have been used as specific epithets, whether 'correct' or synonyms, are included. Names of subspecies, forms and varieties (unless they have been used at some time as specific epithets) are not included, except in a few instances where they have been mentioned in the text.

Present 'correct' specific names are in CAPITALS, synonyms and sub-specific names in *italics*. Page numbers in heavy type (e.g. p. 21) relate to full descriptions, and those in *italics* to illustrations.

[153]

[155]

lacei = *Dionysia lacei*
LACERATA 55
LACINIATA 48, 80
lacteocapitata – subsp. of CAPITATA
LACTIFLORA 39
LAETA 55
lanata = HEUCHERIFOLIA
lancifolia = RUSSEOLA
langkongensis = MALVACEA
LATISECTA 39
LAXIUSCULA 98
lecomtei = FABERI
legendrei – a subsp. of SOULIEI
leimonophylla = VIRGINIS
lepida = FARINOSA var. *denudata*
lepta = APOCLITA
LEPTOPHYLLA 55
leptopoda = STENOCALYX
leucantha – a var. of BEESIANA
leucochnoa = MELANOPS
leucops = RIGIDA
lhasaensis = JAFFREYANA
licentii – a var. of GEMMIFERA
lichiangensis = POLYNEURA
LIMBATA 115
limnoica = *alta*
limprichtii = OVALIFOLIA
listeri Forrest = SINOLISTERI
LISTERI King 49
listeri Pax = SINOLISTERI
LITTLEDALEI 113
littoniana = VIALII
loczii Farrer = ERRATICA
loczii Kanitz = STENOCALYX
LOESENERI 39
longifolia Bieb. = ALGIDA
longifolia Curtis = AURICULATA
LONGIPES 115
longipetiolata = LIMBATA
LONGIPINNATIFIDA 79, 80
longiscapa Komarov = FISTULOSA
longiscapa Ledeb. = INTERMEDIA
longituba = MEMBRANIFOLIA
LUDLOWII 106

luteofarinosa var. *denudata* =
    ALGIDA
LUTEOLA 83

maclareni = LOESENERI
macounii = *arctica*
MACROCARPA Max. 81
macrocarpa Tanaka = *hakusanensis*
macrocarpa Watt – a var. of
    MACROPHYLLA 121
MACROPHYLLA Don 115, 118, **120**,
    121, 124
macrophylla Koch = AURICULATA
macropoda = OVALIFOLIA
maikhaensis = UMBRELLA
MAIREI 102
MALACOIDES 4, 72
MALLOPHYLLA 135
MALVACEA 50, 51
mandarina = SINENSIS
MATSUMURAE Petitm. 80
matsumurae Takeda = MODESTA
MAXIMOWICZII 115, 118, **126**
meeboldii = *moorcroftiana*
meiantha – a subsp. of FORBESII
MELANODONTA 135
MELANOPS 115, **121**, *Pl. XXVI*
melichlora = SPATHULIFOLIA (?)
MEMBRANIFOLIA 80
menziesiana = BELLIDIFOLIA
MERRILLIANA 48
metria = UMBRATILIS
microdonta = SIKKIMENSIS
MICROLOMA 135
micropetala = BELLIDIFOLIA
microstachys = BLATTARIFORMIS
MINKWITZIAE 39
minima Mertens = CUNEIFOLIA
MINOR 115
MINUTISSIMA Jacq. 77
minutissima Klatt = CONCINNA
mirabilis = *Androsace mirabilis*
MISHMIENSIS 115
MIYABEANA 135

[160]